This book is a tool for the theoretical and numerical investigation of nonlinear dynamical systems modelled by means of ordinary differential and difference equations. Special attention is given to the analysis and understanding of chaotic dynamics. The work is divided into two parts: a book, comprising a theoretical overview of the subject matter and a number of applications; and an integrated software program (called DMC) together with its manual.

Although the emphasis is laid on dynamical systems arising from economic motivation, and the applications are derived from these systems, both the text and the program will also be of use to researchers in other fields of study. While it is not intended to be a textbook in mathematics, the mathematical language is of a precision comparable to that used in analogous books for applied scientists. The book first discusses the fundamental concepts and methods of chaos theory, and then applies these theoretical results and the facilities provided by the companion software program to models suggested by economic problems.

Chaotic Dynamics

Chaotic Dynamics

Theory and Applications to Economics

ALFREDO MEDIO

DMC Software

by
GIAMPAOLO GALLO

CAMBRIDGE
UNIVERSITY PRESS

Published by the Press Syndicate of the University of Cambridge
The Pitt Building, Trumpington Street, Cambridge CB2 1RP
40 West 20th Street, New York, NY 10011-4211, USA
10 Stamford Road, Oakleigh, Melbourne 3166, Australia

First published 1992. Reprinted 1993

Printed in Great Britain by
Athenæum Press Ltd, Newcastle upon Tyne

A catalogue record for this book is available from the British Library

Library of Congress cataloguing in publication data

Medio, Alfredo, 1938–
Chaotic dynamics: theory and applications to economics / by
Alfredo Medio; DMC software by Giampaolo Gallo.
p. cm.
Includes index.
ISBN 0-521-39488-0 (hardcover)
1. Statics and dynamics (Social sciences) 2. Economics,
Mathematical. 3. Differential equations, Nonlinear. 4. Chaotic
behavior in systems. I. Title.
HB145.M43 1992
330'.01'51–dc20 91–19382 CIP

ISBN 0 521 39488 0 hardback

To Marji

Contents

Part III: Software

Preface

My interest in nonlinear dynamics was first aroused in the 1970s in Cambridge during the preparation of a Ph.D. thesis on the theory of the business cycle under the supervision of Nicholas Kaldor and Richard Goodwin.

After a few years' diversion due to various circumstances of life, I took up the subject again in earnest when I joined the Department of Economics at the University of Venice in the early 1980s. I soon became convinced that, besides its obvious intellectual appeal, the 'chaos revolution' in dynamical systems theory was greatly relevant to economic theory and was worth a substantial investment of time and effort. A brief visit to Santa Cruz in 1984 and a conversation with Ralph Abraham also convinced me that the intelligent use of numerical simulations had been and would remain an indispensable tool for a good qualitative understanding of complex behaviour of dynamical systems. A turning point in my research was reached during sabbatical leave in Cambridge in 1986–87, where I finally had the leisure necessary to reflect upon, and to begin to understand, new and difficult ideas. In this process, I was greatly stimulated by the seminars and discussions of the King's College Dynamical Systems Group.

Back in Venice, I realized that, no matter how long I tried, my programming proficiency would never reach the minimum standard required for my projects, and that I needed professional help in this area. By lucky coincidence, a young computer wizard, Giampaolo Gallo, was then completing his undergraduate work at Venice and enthusiastically accepted applying his skills to the study of chaos. It was the beginning of a three-year, very fruitful collaboration from which the DMC program has emerged.

In the meantime, my mathematical interests brought me in contact with Sergio Invernizzi of the Mathematics Department of the University of Trieste, who found some of the problems arising in economic dynamics challenging and worth serious mathematical investigation. We had great fun in writing an article together on lags and chaos, and I am indebted to him for innumerable discussions over the last few years from which my understanding of nonlinear dynamics (and therefore the quality of this book) has been greatly enhanced. I sincerely hope that at least some of the n research projects we have been considering will be realized in the not too distant future.

Apart from these general acknowledgements, I wish to thank personally, and also on Giampaolo's behalf, some of the people who, in various capacities, have contributed to this work. In the preparation of the book, I have benefited, in one form or another, from the comments and the criticisms of Giancarlo Benettin, Franco Chersi, Franco Donzelli, Sergio Invernizzi, Gregory King, Marji Lines, Franco Nardini, Giulio Pianigiani, Nicola Rossi, Domenico Sartore, Colin Sparrow and Claudio Tebaldi. I am also grateful to the group of students who attended the series of informal seminars on dynamical systems held in Venice in spring 1990 and contributed many intelligent questions and criticisms. Roberto Perli Traverso deserves special thanks for his help both with the preparation of DMC and the numerical simulations used in the book. Roberto's contribution to specific parts of this book is acknowledged in the relevant chapters below.

For the realization of DMC, Giampaolo and I are indebted to the friends and colleagues of the Computer Centre (Centro di Calcolo) of the University of Venice, and in particular to Francesco Dalla Libera, Alberto Gambi and Otello Martin for the generous allowance of their time and the facilities of the Centre. In this context, we also wish to thank Dr Giovanni Alberini of TecLogic of Padua for his support and help in the preparation of the Macintosh version of DMC.

This book has been prepared by Giampaolo Gallo and myself using the plain format of the TEX typesetting system, together with certain *ad hoc* macros provided by Cambridge University Press. Very valuable help in the technical realization of this book and the DMC Manual has been provided by Geraldine Ludbrook, Germano Galletti and Holly Snapp. Patrick McCartan, John Haslam, Chris Doubleday and Mark Reed of Cambridge University Press have also been very helpful throughout the lengthy realization of this project.

Financial support for the research on which the book is based has been provided, at various times, by the Italian National Council of Research (CNR), the Italian Ministry of Education (later Ministry for Universities and Research) and the Ente Luigi Einaudi. To all these institutions I wish to express my gratitude.

Alfredo Medio

Acknowledgements

For permission to reproduce text figures from a number of publications – with various amounts of alterations – the author would like to thank: Professor Ralph Abraham and Aerial Press for Fig. 9.9; Zeitschrift für Naturforschung and Robert Shaw for Fig. 7.2 (from Shaw, 1980); Physic–Verlag and Professor H.G. Schuster for Figs. 9.2, 9.3, 9.5, 9.6 (from Schuster, 1989); John Wiley & Sons for Figs. 4.2, 4.6, 6.3, 8.1 (from Moon, 1987) and Figs. 2.16, 8.3, 9.7, 9.8 (from Thompson and Stewart, 1986); Springer Verlag and Professors Lichtenberg and Lieberman for Fig. 6.2 (from Lichtenberg and Lieberman, 1983); Springer Verlag and Professor Wiggins for Fig. 2.11 (from Wiggins, 1988); Journal of Mathematical Economics and Professor Invernizzi for Fig. 11.2 (from Invernizzi and Medio 1991).

For permission to reproduce (more or less revised) parts of the articles Medio 1991b, Invernizzi and Medio 1991, and Medio 1991a, the author would also like to thank, respectively, *Journal of Economic Behavior and Organization*; *Journal of Mathematical Economics* (together with Professor Sergio Invernizzi); *Structural Change and Economic Dynamics*.

Part I

Theory

1

General introduction: chaos and economics

1.1 INTRODUCTION

The objective of *Chaotic Dynamics. Theory and Applications to Economics* is to provide a tool for the theoretical and numerical investigation of nonlinear dynamical systems modelled by means of ordinary differential and difference equations, with special attention given to the analysis and understanding of chaotic dynamics. The work is divided into two parts: (i) a book, including a theoretical overview of the subject matter and a number of applications; (ii) an integrated software program called DMC[1] together with its manual. It is hoped that both the text and the program will suit researchers and students in different fields of study, even though the emphasis is laid on, and all applications are derived from, dynamical systems arising from economic motivation.

The potential reader will perhaps wish to raise one or two preliminary questions concerning the tenor of the book and the audience to whom it is addressed. First of all, s/he may wonder how the present work is related to the already voluminous literature on chaos, and what is the level of mathematical sophistication chosen by the author. In general, we have tried to comply with Einstein's much quoted maxim that 'things must be made as simple as possible, but not simpler'. Most interesting results on chaos and nonlinearity are mathematically far from trivial, and the subject could not be treated in a nontechnical fashion, lest the meaning and novelty of these results be lost. Although this is not intended to be a textbook on mathematics, we have tried to keep the mathematical language at the same degree of precision as is used in analogous books for applied scientists. (This is also, by and large, the level of technical sophistication adopted by journals specializing in

---

[1] In the early version DMC stood for Dynamic Model Cruncher and the acronym has been retained.

mathematical economics.) All proofs, however, and many deep questions have been omitted and the interested reader is referred to the original contributions, most of which are to be found as articles scattered in various technical journals.[2]

Moreover, since consistent and universally accepted definitions of the main ideas and methods are still lacking, we have devoted some space and much effort in an attempt to clarify those issues about which confusion is sometimes created by improper or inconsistent terminology or interpretation in economic applications.

A book on nonlinear dynamics and chaos written by a mathematical economist may prompt the reader to ask a second preliminary question: why should economists be mindful of chaos? Or, conversely, why should chaos theorists be mindful of economic problems?

These issues cannot be conclusively decided by any amount of *a priori* reasoning, however competently stated, but will ultimately be settled by the production of (or the failure to produce) hard results which are both mathematically sound and economically relevant. So the appropriate conclusion might be to skip preliminaries altogether and 'get on with the job'. However, sound arguments alone are seldom effective without a certain amount of rhetorical persuasion. This is especially so when new ideas, problems or tools of analysis are put forward, or when certain hitherto separated areas of research are drawn together.

We shall therefore devote a few pages to the introduction of the subject of the book, linking it to certain areas of research in theoretical and empirical economics, and claiming the relevance and potential fruitfulness of the methods discussed.

1.2 AN INTUITIVE DEFINITION OF CHAOS

Although a universally accepted and comprehensive characterization of chaos is still lacking, in this introductory chapter we shall adopt the definition suggested in a recent conference on chaos (Royal Society, London, 1986), namely, 'stochastic behaviour occurring in a deterministic system'. Broadly speaking, a system is said to be deterministic when it comprises no exogenous random variables. On the other hand, the *observable* behaviour of a dynamical system is called stochastic when the transition of the system from one state to another can only be given a probabilistic description as happens for truly random processes, e.g.,

---

[2] Some very good books of readings have recently been published. See, for example, Hao (1984) and Cvitanovic (1984).

the outcome of spinning a roulette wheel. The meaning of these rather intuitive considerations will be made clearer, we hope, in the following pages. Indeed, much of this book is devoted to the discussion of precisely this point.

But how can we reconcile these two apparently opposite modes of behaviour, random and determinate? A full answer to this question would entail the consideration of deep questions of mathematics and probability theory, which cannot be attempted here.[3] Some of the technical details relevant to our discussion will be dealt with in the following chapters. For our present purposes, a few basic nontechnical characterizations of deterministic chaos will suffice.

Let us make the following assumptions:

(a) a certain agency, call it 'news', has a precise knowledge of the rules governing the evolution in time of a dynamical system, over an appropriate variable space;

(b) this space is partitioned into a finite set of equal, non-overlapping, numbered cells, the size of a cell depending on the accuracy with which the position of the system is monitored (this procedure is sometimes called 'coarse-graining');

(c) at certain arbitrarily small time-intervals, we are told by the 'news' the number of the cell where the system then resides.

Let us now pose the following question: is the complete knowledge of which cells the system has occupied in the past sufficient to predict its future course?

For a nonchaotic system, the answer to this question must be affirmative. In this case, the sequence of cell numbers will eventually follow a regular pattern, e.g., it will be periodic. Sooner or later, therefore, when we have accumulated a sufficient number of data, we shall be able to exactly predict the future positions of the system moving from one cell to the next. For nonchaotic systems, therefore, after a certain time, listening to the 'news' becomes totally uninformative.

For a chaotic system, however, no matter how long we accumulate data on the past positions of the system, we cannot accurately predict its transition from the present position to the next one (or to any of the future ones). In this case, we can only make probabilistic forecasts, if any at all. For chaotic systems, therefore, coarse-grained past, however

---

[3] A very nice introductory discussion of this point, which has inspired the next few paragraphs, can be found in Ford (1983).

long, does not determine uniquely and completely coarse-grained future. In other words, the 'news' will continue indefinitely to be a source of additional information on the system.

This characteristic of chaotic systems persists no matter how small the size of the cell in the finite partition, i.e., no matter how accurate the 'news' is. Therefore, for a system following a chaotic orbit, despite exact knowledge of its 'law of motion', the sequence of the cell numbers corresponding to its positions may look random and provide no obvious hint of the underlying deterministic structure, for any finite number of observations of finite precision. The fact that we cannot make exact predictions of the long-term behaviour of chaotic systems does not exclude the possibility of making (more or less accurate) short-run forecasts. The general qualitative nature of those systems can also be investigated by estimating certain statistical invariants, which will be discussed in some detail in the following chapters.[4] However, even statistical predictions may become difficult or impossible if a chaotic system exhibits the so-called 'sensitive dependence on parameters', i.e., if its mode of behaviour changes drastically when the value of a parameter is altered by an arbitrarily small amount. In this case, statistical averages concerning the system may become unstable under variations in parameters, so that the forecaster may have to face a situation that is even worse than that created by sensitive dependence on initial conditions.[5]

1.3 THE DISCOVERY OF CHAOS

Although the existence of deterministic dynamical systems with chaotic, or 'strange', behaviour has been known for over a century, only recently has the scientific community awoken to the fact that chaos is in no way a rare or pathological occurrence. On the contrary, most dynamical systems exhibit chaotic dynamics in the sense just outlined.

This is a disturbing result. In fact, although there is no fixed limit to the degree of accuracy of observation of physical phenomena, we must assume that it is finite, i.e., the observations are necessarily coarse-grained,[6] 'infinite precision' (very much like 'perfect foresight') being only a convenient but possibly misleading abstract concept with no

---

[4] On the prediction of the evolution of chaotic dynamical systems, see, for example, Crutchfield and McNamara (1987); Farmer and Sidorowich (1988).

[5] On the concept of sensitive dependence on initial conditions and on parameters, see Chapters 2 and 9 below.

[6] To this effect one could (but in most cases of practical interest, one need not) make recourse to Heisenberg's Principle of Indeterminacy.

counterpart in physical phenomena. It follows that for the generality of dynamical systems of practical interest, absolute determinism of the 'laws of motion' does not necessarily exclude randomness of behaviour.

Naturally, since any conceivable modelling of a real system must necessarily entail a selection of relevant variables to the exclusion of the rest, the presence of external shocks and the noise that accompanies them can never be completely eliminated. Therefore, in order to be compared to real data, a representation of a real dynamical system will have to include some stochastic elements. However, the 'chaos revolution' of recent years has shown that external noise may not be the only or even the most important source of randomness in the behaviour of real systems, but there may also be an 'intrinsic randomness' owing to the unavoidable fact that we cannot measure physical variables with infinite precision.

The story of the discovery, neglect and revival of deterministic chaotic dynamics, i.e., the dynamics of systems which are both deterministic and stochastic, is scattered among many books and articles, with various degrees of sophistication and originality, and it will only be briefly sketched here.[7]

The possibility that simple, nonlinear, low-dimensional dynamical systems can exhibit very complicated, apparently random behaviour, was already known to Henri Poincaré at the turn of the century. The idea remained (almost) dormant for several decades, in spite of some important contributions to the field, such as those made by Birkhoff in the USA and Cartwright and Littlewood in the UK. In the West it was highlighted in the early sixties by the work of Edmund Lorenz (1963) in the context of a numerical investigation of a model of atmospheric turbulence. Lorenz's results, it is important to note, were made possible by the availability of electronic computers, which allowed the 'step-by-step' computation of differential equations. Basic theoretical results on chaotic dynamics and 'strange' attractors were provided in the 1960s by Smale (1963, 1967) and in the early 1970s by Ruelle and Takens (1971). In the meantime, some very important results were produced in the USSR both in the ergodic and in the geometric theory of chaos (among others by Kolmogorov, Arnold, Anosov, Šilnikov and Sinai), some of which became widely known in the West only with a considerable delay. The subject

---

[7] A very readable, nontechnical report on the story of chaos from an American point of view can be found in Gleick (1987).

has in the recent past experienced a tremendous rate of development and the literature is still growing fast.

Very significant progress has been made mainly in three directions: mathematical investigation of nonlinear dynamical systems; applications of mathematical results to the analysis of models motivated by problems of physical interest;[8] analysis of so-called 'experimental signals', i.e., time series supposedly generated by (unknown) dynamical systems.

Two aspects of chaos theory have been receiving an increasing amount of interest in the last two decades:

(i) chaotic behaviour may be strange, but, as we have already mentioned, it is not rare;

(ii) the chaotic behaviour of increasingly large classes of dynamical systems can be described by a relatively small number of mathematical objects and certain universal properties (e.g., period-doubling sequence of bifurcations) have been discovered which do not seem to depend on the specific systems under investigation, whether these emerge from biology, physics, economics, etc. Furthermore, the modes of behaviour predicted by the analysis and the numerical simulations of theoretical models have often been confirmed by physical experiments in different disciplines, though not yet, or not yet conclusively, in economics.

Thus chaos theory has from the beginning had a rather strong interdisciplinary character, though its universal nature has sometimes been exaggerated and occasionally associated with a scientifically dubious 'dark mystique'.

1.4 CHAOTIC DYNAMICS AND BUSINESS CYCLES

Since many important topics in economics are typically formalized by means of systems of ordinary differential or difference equations, these findings alone should be sufficient to motivate economists' broad interests in chaos theory. But there exists a question, or rather a group of questions, in economics, usually labelled 'business cycles', for which the field of mathematical research under discussion is eminently important.[9]

---

[8] Throughout the book, the word 'physical' does not refer solely to problems of physics, but, more generally, to questions arising from the study of the time evolution of real systems, whether the latter pertain to physics itself, or to other disciplines like biology, economics, etc.

[9] There exist, of course, other areas of research in economics for which chaos theory is, or could be shown to be, very important, e.g., technical progress. We believe, however, that

An appropriate definition of business cycles can be drawn from the classical work of Burns and Mitchell:

Business cycles are a type of fluctuation found in the aggregate activity of nations that organize their work mainly in business enterprises; a cycle consists of expansions occurring at about the same time in many economic activities, followed by similarly general recessions, contractions, and revivals which merge into the expansion phase of the next cycle; this sequence of changes is recurrent but not periodic (1946, p. 3).

For certain aspects, cyclical fluctuations thus described by Burns and Mitchell have been to economists what turbulence has been to physicists: a phenomenon the existence and relevance of which few doubt, but a full understanding of which still escapes us. Business cycles have been a conundrum especially for theoretical economists formed in the tradition of general competitive equilibrium (henceforth GCE) theory.[10]

In its Walrasian formulation, the theory of GCE was historically conceived in order to explain interdependence of economic variables (essentially prices and quantities produced and exchanged) at a given moment in time, rather than their evolution in time. In its atemporal version, with given tastes, technology and distribution of property rights, equilibrium prices and quantities are determined by the condition that all markets 'clear'. That is to say: at the equilibrium prices, supplies and demands – generated by atomistic utility-maximizing consumers subject to budget constraints, and by atomistic profit-maximizing producers subject to technological constraints – are equal in all markets.

The modern, intertemporal (Arrow–Debreu) version of GCE can be viewed as a re-interpretation of the static model, which explicitly takes account of time. In this version, commodities are characterized not only by the usual physical properties, but also by the date of delivery. If we assume that future markets exist for all commodities and for all dates,[11] then an equilibrium can be defined as a set of spot and future prices and a set of quantities of commodities to be delivered now and at future

---

the case of business cycles can best illustrate the role of nonlinear dynamical analysis in general and of chaos theory in particular, especially when we look at it in a historical perspective.

[10] Much of what follows could also be said of the other main strand of equilibrium theory based on the concepts of reproduction and surplus, which is sometimes labelled *classical* or, in its contemporary version, *neo-Ricardian*.

[11] As Kenneth Arrow has recently observed, this is equivalent to the assumption that all individuals correctly anticipate all future prices, which, stochastic disturbances apart, is in turn equivalent to the so-called rational expectations hypothesis (see Anderson, Arrow and Pines (1988), henceforth AAP, p. 276).

dates, such that supplies and demands are equated for all commodities and for all periods of time. Clearly this system implicitly introduces into the picture borrowing and lending as well as interest rates.

The central role played by general equilibrium theory in economic analysis can be explained in terms of a number of different factors some of which are scientifically more significant than others. The most important explanation of this basic feature of economic theory lies perhaps in the logical link which has been established by theoretical welfare economics between competitive equilibria and Pareto optima.[12]

Thus, although GCE is in principle a descriptive theory, it has always been characterized by a strong normative content. Most economists' policy recommendations, and in particular their assessments of the relative desirability of competition and government intervention, have largely been based on propositions concerning GCE. The normative interpretation of the theory, by the way, explains why the existence of multiplicity of equilibria, a perfectly general possibility, has typically been considered by economists as undesirable and often neglected.

All this said, it is certainly hard to find a justification for economists' overwhelming preoccupation with equilibrium states, unless one believes that they provide satisfactory representations of observable configurations of real economies. A necessary condition for equilibria to possess this property is that they may be described as the terminations of actual dynamical processes, i.e., that they are stable under reasonable assumptions concerning agents' off-equilibrium behaviour. Moreover, if transient dynamics have to be neglected, convergence to equilibrium must be sufficiently fast.

The question of what happens if actual prices (or quantities) are not equal to those in equilibrium was, of course, raised at an early stage, but it was never given a thorough and generally accepted answer within the GCE framework. From its inception, the prevailing theory emphasized amplitude-reducing (negative-feedback) behaviour in the economic system. It was (explicitly or implicitly) assumed that competitive equilibrium is stable and that any perturbations will be followed by a swift return to it, with or without short-term damped oscillations.

GCE economists sometimes justify their emphasis, or rather belief, in stabilizing forces by the broad qualitative observation that market

---

[12] For those readers who are not acquainted with economic theory, we shall recall that an allocation of resources among economic agents is called 'optimum in the sense of Pareto' when the position of no agent can be improved (in terms of utility) without worsening that of some other agent.

economies as a whole seem to have a certain degree of self-maintenance (e.g., prices do not explode, most shocks seem to be absorbed rather quickly, etc.). However, this argument confuses the boundedness of the time evolution of a system and its resilience in the face of shocks, which are presumably prerequisites of viability and observability, and the stability of a particular state (equilibrium) which need not be such a prerequisite. On the other hand, the fact that each individual non-equilibrium state of the system is ephemeral (because to that state there correspond arbitrage opportunities that rational agents are bound to exploit) does not mean that a *set* of such non-equilibrium states may not be persistent. In other words, an attractor of a (nonlinear) dynamical system need not be a point.

At any rate, when the stability hypothesis was put to a rigorous test by formulating systems of differential or difference equations incorporating explicit and precise assumptions concerning off-equilibrium behaviour of agents, no compelling general reasons could be found for the equilibrium of the system to be stable, and those sufficient conditions for stability which could be rigorously established were found to correspond to very stringent and economically arbitrary constraints. The rather curious reaction of most economists to these findings, instead of abandoning the hypothesis of stability of equilibrium or thinking again, was to drop the problem altogether, by simply *assuming* that the system is always at equilibrium.

1.5 EQUILIBRIUM AND DISEQUILIBRIUM DYNAMICS

Although the static bias characterizing the original formulations of GCE was essentially maintained in the more elegant and sophisticated modern versions, the intertemporal set-up of GCE does have a built-in dynamical characterization. Intertemporal equilibrium in fact defines certain time paths of prices and quantities (as well as interest rates) which can be (and usually are) called 'equilibrium dynamics'.

To make sense economically, this recursive form of general equilibrium or, as it is sometimes called, 'sequence equilibrium'[13] must presuppose the existence of two types of dynamics, which, for brevity, we shall

---

[13] For simplicity's sake, we ignore here the difference between a situation in which the sequence of equilibrium, dated prices and quantities is fixed at a given instant of time for all the infinite future, and the situation in which those prices and quantities are determined at different subsequent times. The two situations are equivalent only under very stringent assumptions and, strictly speaking, the term 'sequence' only applies to the latter.

label 'short-run' and 'long-run' dynamics. The former type of dynamics controls the reaction of the system to deviations from its temporary equilibrium. Since it is (usually implicitly) assumed that the adjustment mechanism is stabilizing and that convergence takes place very rapidly, in fact instantaneously, 'short-run dynamics' are often hidden. Thus, for example, a stable, infinitely fast adjustment is supposed to guarantee that demands and supplies are always equal in all markets. Mathematically, this situation is depicted by writing a set of algebraic equations corresponding to equilibrium conditions, rather than a set of differential (or difference) equations corresponding to the short-run reaction-mechanisms. The second type of dynamics controls the evolution in time of (temporary) equilibrium states. This, in turn, may or may not lead to a stationary state.[14]

Given the constraints imposed on equilibrium dynamics by the assumptions usually adopted by the prevailing theory (concavity of utility and production functions, constant returns to scale, intertemporally independent tastes and technology, etc.), equilibrium dynamics is not the most promising breeding ground for instability and complexity.[15] However, the observation that real economies do not seem to confirm the predictions of the theory has prompted a number of contributions explicitly aimed at finding characterizations of 'equilibrium dynamics' which may result in oscillatory, or more complex, behaviour of output and prices. The following potential factors of complex behaviour for economies in dynamic equilibrium have been specifically investigated: agents heavily discounting the future; asymmetries of production functions; incomplete markets (of which the overlapping-generations model may be considered a special case); imperfections of markets; intertemporally dependent utility functions. Some interesting and promising results have been produced in the last fifteen or twenty years in this line of research,[16] no doubt stimulated

---

[14] Notice that the assumption that the system is always in (temporary) equilibrium while equilibrium itself is evolving in time may be particularly misleading when the stability properties of short-run dynamics depend on the values of the long-run variables. The assumption of instantaneous adjustment in fact amounts to treating long-run variables as constant parameters with regard to short-run dynamics. However, it can be shown that near bifurcation points, the short-run behaviour of the system for fixed parameters may be very different from the 'true' behaviour which is obtained when the long-run variables are allowed to vary, no matter how slowly (cf. Wiggins, 1990, pp. 384–6)). The relevance of this point to economics can be appreciated by considering economic models characterized by 'relaxation' oscillations, for example Uzawa's (1963) well-known two-sector model.

[15] See, on this and related points, Brock's interesting comments in AAP (1988, pp. 77–97).

[16] This is no place for a comprehensive survey of the literature on these and similar

by the recent developments in nonlinear dynamic analysis. Even though the logical possibility of 'irregular' behaviour in equilibrium dynamics has been shown beyond any possible doubt, one may wonder whether this is the most appropriate setting for explaining observed fluctuations of real economies.

This brings us back to an older tradition of business cycles theory, which is based on the idea that instability and oscillations are essentially due to market failures and which, somewhat misleadingly, is associated with the name of Keynes. *Faute de mieux*, we shall label the approach 'nonlinear, disequilibrium' (NLD) theories of the cycle. In the years that followed the publication of Keynes's *General Theory*, a number of mathematical models of the cycle were produced, most of them sharing the following common characteristics:

(i)   hypotheses derived from broad generalizations of observed 'stylized facts';

(ii)   high level of aggregation;

(iii)   deterministic nature, i.e., the explanations of business cycles were sought in certain properties inherent to the economic system, rather than in the action of external random shocks;

(iv)   nonlinearity[17] of certain basic functional relationships of the system;

(v)   replacement of equilibrium conditions with adjustment mechanisms formulated in terms of (discrete or continuous) lags;

(vi)   quantities adjusting much more rapidly than prices, so that cyclical movements essentially take place at constant prices. (This fact, incidentally, appeared to give some justification to the aggregation of individual quantities into global aggregates);

(vii)   emphasis concentrated on investment volatility, so that the main nonlinearities usually concerned the investment function, often a variation of the 'acceleration theory';

(viii) expectations, often implicit in the adjustment mechanisms, of an adaptive (backward-looking) type.

variations of the basic 'equilibrium dynamics' model, but we shall quote some of the most significant contributions: on the role of 'impatience', see Benhabib and Nishimura (1979), Deneckere and Pelikan (1986) and Boldrin and Montrucchio (1986); on overlapping-generations models, Benhabib and Day (1982), Grandmont (1985) and Reichlin (1986); on imperfectly competitive markets, Woodford (1987); on intertemporally dependent utility, Ryder and Heal (1973).

17 One might as well say 'non-monotonicity', rather than simply 'nonlinearity', since it is the former property which is usually responsible for 'irregular' behaviour. See, on this point, Hirsch (1985).

Examples of such models can be found in the works of Kaldor (1940), Hicks (1950) and Goodwin (1951).[18] Mathematically, the typical result of the early NLD models of economic fluctuations was the appearance (under certain conditions on the parameters) of a limit cycle, which was taken as an idealized description of self-sustained real cycles, with each boom containing the seeds of the following slump and vice versa. This approach to the analysis of business cycles was very popular in the forties and fifties, but its appeal to economists seems to have declined rapidly thereafter and a recent, not hostile textbook of advanced macroeconomics (Blanchard and Fischer, 1987, p. 277) declares it 'largely disappeared'.[19]

The reasons for the crisis of the Keynesian style of theorizing and the associated NLD theories of the cycle are manifold, not all of them perhaps pertaining to scientific reasoning, and a full investigation of this interesting issue is out of the question here. However, there exist certain, partly overlapping fundamental criticisms, raised against this approach mainly by supporters of the rational expectations hypothesis, which are relevant to our discussion and can be briefly summarized thus.

(i)   In Keynesian models, macroeconomic relationships are not derived rigorously from sound 'first principles', by which term economists usually mean the solutions to dynamic optimization problems. In particular, the behaviour of the system in disequilibrium, as well as the various leads and lags associated with it, is purely *ad hoc* and lacks a sound microeconomic underpinning.

(ii)   Agents' expectations, either explicitly modelled, or implicitly derived from the overall structure of the model, are, under most circumstances, incompatible with agents' 'rational' behaviour.

(iii) The NLD approach has been 'refuted' by empirical observation, as time series generated by the relevant models do not agree with available data. For example, so the argument runs, a time series generated by a model characterized by a stable limit cycle will have

---

[18] Kalecki's models of the cycle (see, for example, (1954)) are somewhat different in two respects. They are linear, and they leave a role for stochastic perturbations, at least in some of the numerous versions. In some sense, therefore, they constitute a bridge between the Keynesian models and those of Slutsky and Frisch, on whose contribution we shall have more to say later.

[19] However, the difficulties encountered by the prevailing macroeconomic wisdom in explaining the persistence of economic fluctuations have prompted the addition to the basic neo-classical model of a series of innovations (adjustment costs, lags and leads due to the presence of fixed capital and inventories, staggered contracts, technological lags, etc.) through which some of the ingredients of the NLD models of the cycle were reintroduced through the back door (see Blinder, 1989, p. 106).

a power spectrum exhibiting a sharp peak corresponding to the dominant frequency, plus perhaps a few minor peaks corresponding to subharmonics (see Chapter 5). On the other hand, aggregate time series of actual variables would typically have a broad band power spectrum, often with a predominance of low frequencies (see, for example, Granger and Newbold, 1977). This criticism, by the way, can be (and has been) raised also against 'equilibrium dynamics' models, insofar as they generate regular cycles.

1.6 THE IMPACT OF CHAOS THEORY ON ECONOMICS

Having thus briefly recalled some of the main themes of the economists' long-lasting theoretical debate on economic fluctuations, we may now wonder what the impact on that debate of the recent advances in nonlinear dynamics has been so far.

Some of the effects of the theoretical developments that concern us here are of a general nature and, we believe, will have a positive and lasting influence on economic dynamics in general and the analysis of economic fluctuations in particular, quite independently of the specific views and tenets of the various schools of thought. The discovery of new fundamental results in nonlinear dynamics and their rapid dissemination have given economists active in this field the tools of analysis and the 'vision' necessary to tackle with greater mathematical rigour some of the difficult problems related to instability and oscillations.

A thorough understanding of such concepts as attractors, chaotic dynamics, fractal dimensions, Lyapunov characteristic exponents, symbolic dynamics, etc., as well as their systematic application to the analytical and numerical investigation of economic dynamic models and time series, is an essential link in a chain of scientific progress which will go well beyond the simple study of fluctuations. It is hoped that those new ideas and methods will bring about a radical change of prospect, and prompt that cultural revolution (from a static to a dynamic outlook) which, advocated by Ragnar Frisch in the thirties, has so far been only partially and intermittently realized.

In this context, we would like to stress a few interrelated ideas which are especially relevant. First of all, a necessary, though not sufficient, condition for the occurrence of complex behaviour of a deterministic dynamical system is the presence of nonlinearity. Linear models produce poor idealizations of the dynamics of real systems because their morphology is limited, i.e., they represent only a very limited number of modes of behaviour. It is a known fact that such models cannot properly describe even the simplest nonconstant dynamics of a system, i.e., periodic orbits, let alone chaos. Although the behaviour of certain physical nonlinear systems can be effectively represented by linearized versions through a change of variables, detrending, etc., there are cases in which the price for the simplification of the analysis obtained by means of a linear representation is paid by a 'loss of physics', namely by the obfuscation of the essential dynamical properties of the real phenomenon the model is meant to capture. Only within a nonlinear approach can one fully realize that equilibrium states, stable or unstable, and even periodic orbits, are rather special configurations of a 'dynamical universe' which is much richer than the rather dull subset revealed by the linear approach.

Secondly, instability of equilibrium does not necessarily imply the 'explosion' of the system (and therefore the 'badness' of the relevant model). In fact instability often opens the way to more complicated and interesting modes of behaviour and above all to intrinsic stochasticity, i.e., chaos.

Finally, for linear dynamical systems, complexity of behaviour increases very slowly with dimension (i.e., the number of state variables), and the degree of complexity required for the application of statistical methods to their study only appears as the dimension becomes very large. In contrast, for nonlinear systems the complexity of solutions exhibits a 'leap' each time their dimension is increased by one unit. This is evident and well known for systems of differential equations of dimension one, two and three, and for difference equations of dimension one and two, although the evolution of complexity for higher dimensions is still not well understood.[20] Even low-dimensional nonlinear systems may have solutions so complex as to justify the application of statistical methods, some of which will be discussed in detail in the following chapters.

These ideas may sound mathematically trivial, but they have perhaps for the first time been taken seriously by the economic profession. Indeed they run against some deeply entrenched (although seldom made explicit) theoretical biases which, somewhat loosely, we would label 'static prejudices'. On the one hand, the belief that equilibrium is strongly stable

---

[20] This point is particularly relevant to economics where the vast majority of (continuous-time) macrodynamic models are described by systems of differential equations of order one or two, allegedly for simplicity's sake. Seldom is the reader cautioned that this simplification may be severely misleading, as it automatically excludes from the potential results of the analysis any attractor more complicated than a fixed point (dimension one) or a limit cycle (dimension two).

prompted economists to take a local, reduced view of dynamics, within which linear approximation holds. On the other hand, a linear approach suggests the idea that unstable equilibria are associated with explosive, unrealistic dynamics and are therefore not suitable idealizations of real systems and should be discarded.

Apart from these general considerations, a specific result, i.e., the rediscovery of stochastic behaviour of purely deterministic systems, may have a more direct, and non-neutral effect on the current disputes among proponents of different theories of economic fluctuations, and reopen issues which were thought to have been definitively resolved. On the one hand, the perfect foresight–rational expectations hypothesis of standard equilibrium dynamics clashes with the results showing that that the outcome of the dynamics may well be chaotic, and therefore essentially unpredictable in the sense discussed above.

On the other hand, and for exactly the same reason, the criticism that NLD theories of the cycle allow for permanently unexploited profit opportunities and therefore imply economic agents' irrational behaviour, may not apply in the case of chaotic systems. This point can best be appreciated by making reference to the original formulation of the rational expectations hypothesis. In Muth's own words, 'the expectations of firms (or, more generally, the subjective probability distribution of outcomes) tend to be distributed, for the same information set, about the prediction of the theory (or the "objective" probability distributions of outcomes)'.[21] If expectations were not rational in the sense defined above – so Muth's argument runs – 'there would be opportunities for economists in commodity speculation, running a firm, or selling the information to the present owners'.[22] This argument does have some validity if the outcomes of the dynamical system under consideration can be accurately predicted once the 'true' model is known, e.g., if the outcome is periodic. In this case, if agents are 'rational', fluctuations can be explained only by exogenous random shocks.[23] However, if the theory implies chaotic (i.e., unpredictable) dynamics of the system, the rational expectations argument loses much of its strength and non-optimizing rules of behaviour – such as adaptive reaction mechanisms of the kind

---

[21] Muth (1961, p. 316).

[22] *Ibid.*, p. 330.

[23] Even when the outcome is periodic, however, if the periodicity is long and the time path very complicated, one may question the idea that well-informed real economic agents would actually forecast it correctly: all the more so if the outcome is quasi periodic, possibly with a large number of incommensurable frequencies.

assumed by the NLD models, or 'bounded rationality' *à la* Simon – might not be as irrational as they may seem at first sight. At any rate, the presence of chaos makes the hypothesis of costless information and the infinitely powerful learning (and calculating) ability of economic agents, implicit in the perfect foresight–rational expectations hypothesis, much harder to accept.

The main difficulty here is to define a theory which satisfies two requisites:

(i)   Economic agents, when facing chaotic dynamics of the system, recognize the limitations of their predictive ability and devise rules of action appropriate to the situation. The theory should define these rules assuming rationality, but not the super-human capabilities required to derive exact predictions of the future course of the economy from the precise knowledge of the 'true' model of the economy, in the absence of random shocks.

(ii)   These same rules of action, together with the other characterizations of the dynamical system in question, generate the same chaotic solutions faced by the agents.

Naturally, we do not yet have any theory satisfying these conditions, but the problem is now posed with greater clarity than before.[24]

1.7 CHAOS THEORY AND STATISTICAL INFERENCE

We have thus far concentrated on the theoretical, purely deductive aspects of the analysis of business cycles. But chaos theory may become equally relevant to the empirically oriented, inductive side of analysis. The latter has historically come to coincide with the econometric approach to business cycles the origins of which may be traced back to the seminal ideas of Slutsky (1927), Yule (1927) and Frisch (1933),[25] and was later developed, and given the status of orthodoxy, by the works of the Cowles Commission in the 1940s and 1950s.

The fundamental idea of the econometric approach is the distinction between impulse and propagation mechanisms. In the typical version of this approach, serially uncorrelated shocks affect the relevant variables

---

[24] See Arrow's comments in AAP (1988, p. 280). The issue of chaos and agents' rationality is also discussed in a very interesting paper by Kelsey (1988).

[25] As was recently pointed out (see Lines, 1991), the position of Frisch differed from that of Slutsky in that the former, but not the latter, stressed the idea that the deterministic part of the economic system has oscillatory, though damped behaviour. The role of exogenous shocks is consequently different in these two authors' works.

through distributed lags (the propagation mechanism), leading to serially correlated fluctuations in the variables themselves.[26] As Slutsky showed, even simple linear non-oscillatory propagation mechanisms, when excited by random, structureless shocks, can produce output sequences which are qualitatively similar to certain observed macroeconomic cycles. As a matter of fact, the vast majority of macroeconomic applications of this approach to business cycles consists so far of linear deterministic systems coupled with additive shocks of the white-noise type.

Although the conventional approach has contributed to the improvement of our quantitative knowledge of business cycles, many empirical phenomena remain which are not well covered by it. Furthermore, the ability of that approach to provide an explanation of business cycles can be (and has been) called in question. For one thing, in most conventional stochastic models, the impulse mechanism consists typically of an *ad hoc*, uninformative random factor. As Hicks observed long ago (1950), to explain fluctuations by means of random shocks is tantamount to a confession of ignorance. This is all the more true when the propagation mechanism is non-oscillatory or, if oscillatory, when it is strongly damped. Secondly, a convincing theory of the propagation mechanism is still lacking, despite substantial research and very many alternative hypotheses. Thirdly, it is now well understood that nonlinear, purely deterministic dynamical systems can generate solutions which look as 'irregular' as actual economic time series, and whose autocorrelation functions and power spectra are very similar to those generated by stochastic models consisting of a stable deterministic linear part and an i.i.d. random element.

In this context, the recent work on chaotic attractors has not only provided new tools, thus enabling economists to produce deeper and more sophisticated analyses of deductive theories of fluctuations, but it has also suggested new avenues for nonparametric dynamical inference.[27]

The main finding here is the possibility of recovering information about an unknown model of indefinitely large dimension through the observation of (possibly univariate) data paths, hypothetically produced by that

---

[26] For completeness' sake, among the impulse-propagation models of the cycle, one should distinguish between those in which random external events affect economic 'fundamentals' (essentially, tastes and technology), and those in which those events directly change only agents' expectations. The latter case has been extensively studied in recent years in the economic literature under the label 'sunspots'.

[27] An extended and very interesting discussion of some of the points touched upon in the following pages can be found in Barnett and Choi (1989).

model. The theorems contemporaneously published by Takens (1981) and Mañé (1981) have provided the vital link between the dynamics of the unknown 'true' system in the state space (possibly of very high, or even infinite dimension), and the dynamics, defined in a 'pseudo-phase space', of a system reconstructed from the observed data. When the 'true' system is dissipative and converges to an attractor, its essential asymptotic properties – geometrical, as well as statistical – can be derived from the knowledge of the reconstructed attractor. For one thing, the latter can be shown to be equivalent, in a properly defined manner, to the attractor of the 'true' system. Furthermore, if dissipation is strong (in the sense that, under the action of the dynamical system, a set of initial conditions shrinks in all but a few directions), the resulting attractor may be low-dimensional, thus permitting a simplified, but correct, representation of a potentially very large system.

Notice that the inference techniques in question do not depend on the user's views concerning the 'true' model of the system, which may well remain unknown forever. Moreover, the method in question while providing the means for deriving information about the asymptotic structure of the unknown system from the available data, at the same time effectively realizes a form of unbiased aggregation of the system. In fact, the method can be looked at as a way of estimating the minimum degree of complexity (in the sense of dimensionality) of the structure of the system under investigation. This is particularly important in economics where the sample size of data is usually rather small and we need to 'economize' on the size of the model.

When the reconstructed attractor is shown to be chaotic, we can conclude that the (nonlinear deterministic) dynamics of the 'true' system are 'irregular' and essentially unpredictable as would be the case for the (linear stochastic) models studied within the conventional approach. There is a basic difference, however, between the two alternative representations: the white-noise shocks, usually postulated by the conventional econometric analysis of fluctuations, are exogenous and uninformative. In contrast, the signal resulting from a deterministic chaotic system is typically nonwhite and informative. By this we mean to say that, from the study of the dynamics of the system along the chaotic attractor, we can obtain better estimates of its short-run evolution and derive relevant qualitative and quantitative information concerning its geometrical and statistical properties, some instances of which are the subject of the following chapters (e.g., fractal dimension, average rate of divergence of nearby trajectories, average rate of information generation, embedded periodicities, etc.).

Naturally the method in question is very recent, and the applications are few (albeit rapidly expanding), and there still exist theoretical as well as computational difficulties, some of which we shall discuss in Chapters 10 and 14. Here, we shall briefly mention just two of them. A first question concerns the quality and quantity of data. In fact the method we have just presented seems to work well only when data available are abundant (and dense) and 'clean' from external noise. Lack of data is a problem particularly felt in economics, where time series are often rather short, and when they cover a long period of time we cannot be sure that they pertain to the same system. Concerning the second difficulty – noise – we agree with Barnett and Choi (1989, p. 152) that economic data are not inherently noisier than those of physical sciences. It is also true that, for the method under debate, noise deriving from the simultaneity bias created by endogeneity is not a problem, since we do not have to choose the dimension of the system *a priori*; we only have to fix a lower bound for the reconstructed attractor based on a guess on the dimension of the 'true' one. All this notwithstanding, if the white noise generated by measurement errors and improper aggregation is large *vis-à-vis* the signal, the latter will be overwhelmed by the former and we shall be unable to recover it. Several methods exist to tackle this difficulty as well as to distinguish deterministic and random components of time series. Some of these methods will be discussed in Chapters 10 and 14. We shall also present a relatively lesser known method in economics, which, in our opinion, is particularly promising, and apply it to certain financial data.

A second question arises from the fact that, after having successfully reconstructed the attractor of a system, and having become convinced that the data investigated are actually generated by a nonlinear system whose attractor has a dimension sufficiently small to be tractable, we still lack a model of the system which makes sense in terms of the actual real-world variables and which may be used, for example, for policy recommendations. To complete the inference procedure, we must presumably make recourse, on the one hand, to some *a priori* theory in which we believe and, on the other, we must rely on traditional statistical methods. Indeed only theory can suggest which of the infinitely many possible representations of the 'true' system we should choose, whereas parameter fitting and the various qualitative statistical tests on the candidate model will help to assess its 'goodness'.[28]

We suspect that the latter comments will sound rather familiar to econometricians involved in some recent debates on (sophisticated versions of) old questions, such as 'measurement without theory, theory without statistical inference, econometrics as a bridge between theory and observation' and so forth.[29]

The issue is of course still open and its relevance is by no means only theoretical. The choice between a nonlinear deterministic and a linear stochastic representation of business cycles has some very important policy implications. In the former case, fluctuations are endogenous, i.e., they depend on the structure of the system and in particular on the nonlinearities that characterize it. Therefore, authors taking this view will recommend anti-cyclical policies which affect the crucial parameters of the system, or even alter its basic set-up. On the opposite side, those who believe that the source of fluctuations lies in external random shocks will conclude that intervention is at best irrelevant and, at worst, positively counterproductive.

1.8 A FEW CAVEATS

The optimistic assessment of the potentialities of a research strategy based on an extended application of nonlinear dynamic analysis to economics implicit in the preceding pages, should be accompanied by a few sobering comments concerning the difficulties and limitations still facing the realization of this strategy.

It is fair to say that, in spite of the considerable progress made so far in this field, the majority of the economics profession, largely on an *a priori* ground, remains sceptical, whereas some dynamical system theorists doubt that economics will ever become a promising area of application for their results. But then the majorities of academic professions are seldom right when evaluating the prospects of deeply innovative changes of methods or themes in their fields of research!

One important difficulty here is connected with the lack of common tradition and language between economists and applied mathematicians. This statement may seem surprising in view of the almost complete mathematization of modern economic theory and the rather high level of

---

[28] Some very interesting comments and suggestions concerning different methods of nonlinear modelling can be found in Farmer and Sidorovich's article in AAP (1988, pp. 99–113).

[29] See, for example, the very interesting collection of essays edited by Granger (1990).

mathematical sophistication reached by theoretical economists. However, if one looks more closely at the situation in the areas of research which concern us here, things do not appear as satisfactory as they may seem at first sight.

First of all, virtually all important results in chaos theory which are not purely mathematical, derive from the analysis of models motivated by physical, biological or chemical, but not economic problems. The counterpart of this fact is that many (perhaps most) applications to economics of chaos theory appear to be rather mechanical translations of a small subset of known results, with a varying degree of value added and *ad hoccery*. (Take an aggregate discrete-time economic model, find the conditions for a unimodal nonlinearity to exist, make it depend on a control parameter, and, abracadabra, you can avail yourself at no extra cost of all the formidable literature on non-invertible maps, and show the existence of periodic orbits, chaos and everything else that goes with it.)

A special, and especially unsatisfactory problem concerns the use of numerical simulations in the investigation (and generation) of theoretical models. There is a general consensus among dynamical system theorists that the formidable progress experienced in this field in recent years could not have been possible without the availability of modern computing facilities and their creative use. Many, perhaps even most, analytical results in this field were anticipated by numerical findings, and many important ideas were first suggested by the computer-assisted analysis and visual inspection of the orbits of models, which were either derived from physical motivation, or built *ad hoc*. Furthermore, without the imaginative use of computers, a proper qualitative understanding of complex behaviour of dynamical systems would have been (and will remain) impossible. And Poincaré has taught us that the qualitative[30] understanding of such systems is often all we can obtain, or all we are interested in obtaining.

In theoretical economics, on the contrary, there remains a rather obstinate prejudice against simulations. Numerical results are considered by many as a weak surrogate for exact, rigorous propositions. Often, perfectly general, but from a practical point of view perfectly irrelevant theorems are deemed superior to meaningful and interesting numerical results. In actual fact, simulations are not second-rate substitutes for

---

[30] René Thom (1977, pp. 4–5) has very effectively reminded us that 'qualitative' is a different concept from 'approximate' (i.e., poor quantitative).

theorems, but something different, their theoretical status being the same as that of experiments in the physical sciences. Numerical simulations by themselves do not and cannot prove anything in a mathematical sense but, when well conceived and rigorously performed, they constitute an indispensable complement to theoretical results in any applied science. Occasionally, computations may be part of the logical structure of a proof. More often they provide qualitative information about problems which are well understood theoretically, suggest new theoretical results, or reveal weaknesses or limitations of the old ones. These considerations are best summarized by quoting one of the founding-fathers of computer science, John von Neumann, who in 1945 prophetically stated: 'Really efficient high-speed computing devices may, in the field of nonlinear partial differential equations as well as in many other fields which are now difficult or entirely denied of access, provide us with those heuristic hints which are so needed in all parts of mathematics for genuine progress.'[31]

The message from these comments is that, in order to maximize their fruitfulness for economics, the new analytical as well as numerical methods of nonlinear dynamics must become part of the standard tool-box of economic theorists. This will permit economists to pursue a greater degree of independence from mathematicians and the physical scientists, creatively adapting mathematical tools to their models rather than vice versa. Moreover, only a first-hand study of those methods and a sustained effort in applying them with originality to concrete economic problems will allow us to distinguish between what is of purely mathematical significance and what is also physically, i.e. economically, relevant. For this purpose, numerical computation will again provide powerful tools and many valuable insights. This process of assimilation of mathematical ideas and techniques occurred in the past, with remarkable success, for other areas of mathematics (e.g., calculus, matrix algebra), and it will hopefully take place for nonlinear dynamics as well.

This is of course a long-run goal and, even if it is prompted by early, encouraging results, its attainment will take rather a long time. It is the aim of the present work to provide some contribution in this endeavour.

1.9 THE CONTENT OF THE BOOK

As we have already said, this work comprises a book and accompanying software. The organization and detail of the software program are

---

[31] Quoted in Lax (1986, p. 571).

described in the manual appended to the book (Chapter 15). It is a menu-driven, integrated program providing the most commonly used computing facilities for the analysis of ordinary differential and difference equations of virtually any dimension and degree of complexity. As it is extremely user-friendly, the program can be used even with little or no knowledge of the underlying theoretical principles and theorems. However, we do not recommend such a 'blind' use of numerical simulations: such use can only be of limited help as a first, easy-going contact with a subject which is difficult and which cannot be learned without a substantial investment of time and effort.

As indicated earlier in this chapter, the theoretical part of the work is concentrated on the analysis, detection and characterization of chaotic (or strange) attractors in systems of differential or difference equations (maps). Our interest is not so much in the purely mathematical aspects of theoretical results as in their applications to physically (economically) motivated problems, with a view to a better understanding of the dynamic behaviour of the resulting models.

The first part of the book continues with a chapter (Chapter 2) in which the basic mathematical ideas and concepts relevant to the following analysis are briefly described and some basic results are recalled. We have been guided here by the principle that those concepts and results used in the following chapters which cannot be taken as known by the 'general public' of economists, should be discussed. Nevertheless, given the tenor of the book, we have had to be selective and no doubt others may have chosen the topics differently. The very short Chapter 3 contains a concise, informal user's guide to the book and program.

Chapters 4 to 8 are devoted to the characterizations of chaotic dynamics. Although these chapters can be read independently from any use of the software program, the selection of the topics and their treatment have been made with a view to providing a theoretical underpinning to computer-assisted investigations. The topics comprise Poincaré maps (Chapter 4); power spectrum analysis (Chapter 5); Lyapunov characteristic exponents (Chapter 6); fractal dimensions (Chapter 7); symbolic dynamics and horseshoes (Chapter 8); routes to chaos (Chapter 9); and experimental signal analysis (Chapter 10).

The second part of the book contains some economic applications of the theoretical results contained in Part I and of the facilities provided by the software program. We would like at this point to state clearly that these models have not been chosen because, in the authors' minds, they represent the 'best' available dynamic models in economics (whatever this may mean). Rather, the choice has been dictated by a pedagogical intent and, more specifically, by the following requirements: (i) to present a portfolio of economic formalizations suggested by different problems or schools of thought; (ii) to allow as wide as possible an application of the various techniques, analytical as well as numerical, discussed in the other parts of the book; and (iii) to include analyses and simulations which are as original as possible, and which are at any rate independently generated, rather than ready-made results already available in the literature. More specifically: Chapter 11 contains a discussion of the class of models (by far the most common in the economic literature) which, although differently motivated economically, are all represented by a one-dimensional, discrete-time model characterized by a 'one-hump' nonlinearity. We concentrate our attention on the lag structure common to all those models and, in so doing, we discuss the question of discrete versus continuous-time characterization of dynamic models in economics. Chapter 12 considers an instance of 'equilibrium dynamics', in the form of an overlapping-generations model with production, leading to a two-dimensional discrete-time system. Chapter 13 discusses a continuous-time three-dimensional model motivated by a theory of inventory cycle of the NLD (or Keynesian) type discussed above. Finally, Chapter 14 deals with the question of statistical inference based on the concept of the 'reconstructed' strange attractor already discussed in Chapter 10. The main theoretical aspects and some of the recent contributions to this area are discussed and a relatively new method is applied to a time series generated by financial markets.

2

Basic mathematical concepts

2.1 VECTOR FIELDS, FLOWS AND MAPS

When we begin the investigation of a given dynamical system, usually the first step is to represent it as a mathematical model, based on some *a priori* knowledge of its structure.

In this book, we shall concentrate our attention on systems of ordinary differential equations (o.d.e.) or of difference equations. Typically a system of o.d.e. will be written as[1]

$$\dot{x} = f(x), \qquad x \in \mathbb{R}^n \tag{2.1}$$

where $f: U \to \mathbb{R}^n$ with U an open subset of $\mathbb{R}^n$ and $\dot{x} \equiv dx/dt$. The vector x denotes the physical (economic) variables to be studied, or some appropriate transformations of them; $t \in \mathbb{R}$ indicates time. In this case, the space $\mathbb{R}^n$ of dependent variables is referred to as *phase space* or *state space*, while $\mathbb{R}^n \times \mathbb{R}$ is called the space of *motions*.[2]

Equation (2.1) is often referred to as a *vector field*, since a solution of (2.1) at each point x is a curve in $\mathbb{R}^n$, whose velocity vector is given by $f(x)$.

A solution of (2.1) is a function

$$\phi: I \to \mathbb{R}^n$$

where I is an interval in $\mathbb{R}$ (in economic applications, typically $I =$

---

[1] Systems of type (2.1), in which f does not depend directly on t are called *autonomous*. If f does depend on t directly, we shall write

$$\dot{x} = f(x, t), \qquad (x, t) \in \mathbb{R}^n \times \mathbb{R} \tag{2.2}$$

and $f: U \to \mathbb{R}^n$ with U an open subset of $\mathbb{R}^n \times \mathbb{R}$. Equations of the type (2.2) are called *non-autonomous*. In economics they are used, for example, to investigate technical progress.

[2] In what follows, we shall encounter and discuss cases in which the structure of the phase space is more general than $\mathbb{R}^n$ (e.g., toroidal phase spaces). For simplicity's sake, however, in this general introduction we take phase spaces to be open subsets of $\mathbb{R}^n$.

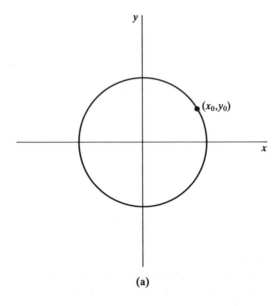

(a)

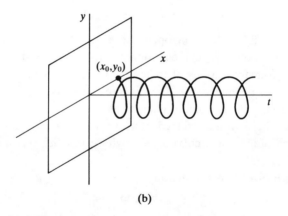

(b)

Fig. 2.1. (a) Orbit through (x_0, y_0) in the (x, y) phase plane; (b) trajectory through (x_0, y_0) at $t = 0$ in the (x, y, t) space.

$[0, +\infty))$, such that ϕ is differentiable on I, $[\phi(t)] \in U$ for all $t \in I$, and

$$\dot{\phi}(t) = f[\phi(t)], \qquad \text{for all } t \in I.$$

The set $\{\phi(t) \mid t \in I\}$ is the *orbit* of ϕ: it is contained in the phase space; the set $\{(t, \phi(t)) \mid t \in I\}$ is the *trajectory* of ϕ: it is contained in the space of *motions*. However, in applications, the terms 'orbit' and 'trajectory' are often used as synonyms. If we wish to indicate the dependence on initial conditions explictly, then a solution of (2.1) passing through the point x_0 at time t_0 is denoted by

$$\phi(t, t_0, x_0),$$

(if t_0 is equal to zero it can be omitted). For a solution $\phi(t, x_0)$ to exist, continuity of f is sufficient. For such a solution to be unique, it is sufficient that f be at least C^1 in U.[3]

We can also think of solutions of o.d.e. in a slightly different manner, which is now becoming prevalent in dynamical system theory and will be very helpful for understanding some of the concepts discussed in the following chapters. Consider that, for autonomous vector fields, time-translated solutions remain solutions, i.e., if $\phi(t)$ is a solution of (2.1), $\phi(t + \tau)$ is also a solution for any $\tau \in \mathbb{R}$. Then, taking t as a fixed parameter, we can write $\phi(t, x) \equiv \phi_t(x)$, where

$$\phi_t(x) : U \to \mathbb{R}^n.$$

Thus, the totality of solutions of (2.1) can be represented by a one-parameter family of maps[4] of the phase space onto itself. In order to study the dynamics of system (2.1) in its phase space under the action of the flow ϕ_t, we will have to investigate the nature of the iterates of points in U under ϕ_t for all $t \in I$.

The map ϕ_t is called a *phase flow* or, for short, a *flow* generated by the vector field f, by analogy with fluid flow where we think of the time evolution as a streamline.

If time t is allowed to take only discrete values, the problem may be formulated as

$$x_{n+1} = G(x_n), \qquad n = 0, 1, 2, \ldots, \tag{2.3}$$

or

$$x \mapsto G(x), \tag{2.4}$$

where $G = \phi_\tau$ and τ is the value of the parameter t.

Dynamical systems like (2.3) or (2.4) are called *discrete-time* (or, simply, *discrete*) *dynamical systems*, since the sequences of values of the state

---

[3] In fact we only need f to be (locally) Lipschitz, i.e., $|f(y) - f(x)| \leq K|x - y|$ for some constant $K < \infty$. Notice that by 'a C^k function' we mean a function which is k-times differentiable.

[4] Roughly speaking, a map is a function which is used to describe the dynamics of a system in a certain space. The term *mapping* is also used. Notice that, if f in (2.1) is C^k in $U(k \geq 1)$, the map ϕ_t is also C^k and it is invertible with C^k inverse. Maps having this property are called *diffeomorphisms*. If $k = 0$, the term *homeomorphism* is used.

variables that they generate correspond to discrete values of the independent variable t. This is not the only way of deriving a discrete-time map from a continuous-time dynamical system. A more important one exists, the Poincaré map, which will be discussed in Chapter 4. Notice that, if a map is derived from a continuous flow by means of a Poincaré surface of section, τ *need not be constant*, but will generally be a function of x.[5]

Not all maps, however, can be conceived as derived from flows generated by systems of o.d.e. like (2.1). Such *flow maps* must be *orientation-preserving*. For a map $\phi_t(x)$ defined above, the latter property holds if and only if

$$\det[D_x\phi_t(x)] > 0, \qquad \forall\, x \in V,$$

where $D_x\phi_t = [\partial\phi_t^i/\partial x_i]$ denotes the matrix of partial derivatives.[6]

Of course, whenever the problem at hand allows it, discrete dynamical systems may be studied in their own right and we might, more generally, study problems resulting in non-invertible maps. Notice that, in what follows, we shall use the term 'flow' broadly to denote continuous-time dynamical systems, and the term 'map' to denote discrete-time systems, whether or not the latter are derived from vector fields.

For a system of ordinary differential equations like (2.1), a general solution $\phi(t, x_0)$ can seldom be written explicitly as a 'combination' of known functions (powers, exponentials, logarithms, sines, cosines, etc.), or in the form of converging power series.

Unfortunately, explicit general solutions are available only for special cases, namely:

(i)   linear systems with constant coefficients;
(ii)   *one-dimensional* systems (i.e., those for which $n = 1$);
(iii) a small number of *very special* nonlinear differential equations of order greater than one.

One-dimensional nonlinear systems of o.d.e., as well as linear dynamic systems, are not very interesting, in the sense that their behaviour is morphologically rather limited and they cannot be used to represent effectively complex dynamics, which are the main object of this book. However, linear theory[7] is important in the analysis of nonlinear systems

---

[5] This point must be stressed since it is regularly neglected in economic applications, where the parameter τ is usually taken as a constant independent of x. A discussion of the relevance of this question to economic applications can be found in Chapter 11.

[6] For a proof of this statement, see Wiggins (1988, p. 74).

[7] There exist several excellent treatises on the theory of linear dynamic systems. See, for example, Kaplan (1962). A good survey of applications to economics can be found in Murata (1977).

since it can be employed – as we shall see in the sequel – to investigate qualitatively their *local* behaviour, i.e., their behaviour in an arbitrarily small neighbourhood of a single point or of a periodic orbit. This is particularly important in stability analysis and in the study of (local) bifurcations.

All this said, the generality of nonlinear systems which are studied in applications escape full analytical investigation. In this case, the short-run dynamics of individual trajectories can usually be described with sufficient accuracy by means of straightforward numerical integration. In applications, however, and specifically in economic ones, we are often concerned not with short-term properties of individual solutions, but with the global qualitative properties of bundles of solutions which start from certain practically relevant subsets of initial conditions. Those global properties, however, can only be investigated in relation to trajectories which are somehow recurrent, i.e., broadly speaking, those trajectories which come back again and again to any part of the phase space which they once visited. In what follows, therefore, we concentrate mainly on that part of the phase space of a system which corresponds to recurrent trajectories in the sense just indicated, which will be made more precise below. Even thus restricted, no comprehensive global theory of nonlinear systems exists as yet, and possibly will never exist. In our opinion, the best research strategy that can be employed at the moment is a combination of analytical and numerical investigation, the latter playing very much the same role as experiments do in natural sciences. This book is in fact intended to provide the reader with some basic tools to carry out such a combined strategy.

Two further considerations should be added at this point. First of all, since an explicit solution for the flow generated by vector field (2.1) is in general out of the question, it is impossible to obtain an exact definition of the maps derived from the flow itself (including of course the Poincaré maps). We shall see how valuable information can be obtained in this case too, by means of approximation and/or numerical computation.

Secondly, analogous difficulties arise when dynamical systems are represented by means of nonlinear maps directly (i.e., without deriving them through a Poincaré section). In this case, too, explicit solutions are generally available only for linear systems. In fact, not even one-dimensional nonlinear maps can usually be given an explicit solution, although, as we shall see (Chapter 9 below), for certain classes of such maps we can obtain a fairly complete qualitative analysis of their dynamics.

In deciding whether to represent a dynamical system in a continuous-time form such as (2.1), or in a discrete-time one such as (2.3) or (2.4), we should keep in mind that the time needed to iterate maps is smaller by a large factor (typically of the order of 10^3) than that needed to integrate differential equations numerically. This advantage, however, may be more than offset by greater analytical difficulties encountered when, in order to provide a satisfactory mathematical representation of the problem at hand, high-dimensional maps are required.

A final important remark should be made at this point. Whereas orbits of ordinary differential equations are continuous *curves*,[8] orbits of maps are *discrete sets of points*. This has a number of important consequences which will become progressively evident in the course of our analysis. Here we should like to mention only one of these consequences, which can be appreciated intuitively. The two basic properties that, under standard assumptions, a solution of autonomous o.d.e. is a continuous curve in the phase space and that only one solution passes through each point in that space, drastically constrain the orbit structure in one- and two-dimensional systems of o.d.e. In the first case, we can only have fixed points and orbits leading to (or away from) them; in the second case, nothing more complex than a limit cycle can occur in general. The restriction does not apply to maps, however, and we shall see that simple, one-dimensional nonlinear maps can generate astonishingly complex orbits.

2.2 CONSERVATIVE AND DISSIPATIVE SYSTEMS

Dynamical systems, whether of a continuous or of a discrete type, can be classified into *conservative* and *dissipative* systems.

A system is said to be conservative if volumes in phase space are conserved by time evolution. Such systems are characterized by the existence of at least one continuous, single-valued, analytic function of the state variables, which is constant (but not identically constant) under the action of the flow (or the map). This constant may sometimes be given an explicit physical interpretation in terms of the problem studied, such as energy. Formally, we may say that the flow associated with an autonomous system of o.d.e.

$$\dot{x} = f(x), \qquad x \in \mathbb{R}^n \tag{2.5}$$

---

[8] We omit here for simplicity's sake the distinction between 'curves' (as functions) and 'images of curves' (as sets of points).

preserves volumes locally if the so-called *Lie derivative* (or *divergence*) is zero, i.e., we have

$$\sum_{i=1}^{n} \frac{\partial f_i}{\partial x_i} = 0.$$

Analogously, a map $x \mapsto G(x)$ is said to preserve volumes in the state space if we have

$$|\det D_x G(x)| = 1.$$

From the fact that in conservative systems volumes remain constant under the flow (or map), we may deduce that those systems cannot have attracting regions in the phase space, i.e., there can never be asymptotically stable fixed points, or limit cycles, or strange attractors. However, the presence of regions in the phase space characterized by very complex or (suitably defined) chaotic behaviour has long since been detected in conservative (specifically in Hamiltonian[9]) systems.

Since strange attractors (to be defined later) are the main object of our investigation and conservative systems are relatively rare in economic applications, we shall not pursue their general study here, and we shall limit ourselves to briefly mentioning two interesting cases.

The first of these is the well-known model developed by Richard Goodwin (1967), who applied certain equations previously studied by the mathematicians Alfred Lotka and Vito Volterra to a classical problem of cyclical economic growth. Mathematically, the LVG (Lotka–Volterra–Goodwin) model has the following form:

$$\dot{x} = x(a - by),$$

$$\dot{y} = -y(c - dx),$$

where $x, y \in \mathbb{R}$ and a, b, c, d are real positive constants. In the Goodwin

---

[9] A continuous-time (autonomous) system of ordinary differential equations

$$\dot{x} = f(x)$$

where $x = (k, q)$, $k, q \in \mathbb{R}^n$ is said to be *Hamiltonian* if it is possible to define a continuous function $H(k, q) : \mathbb{R}^{2n} \to \mathbb{R}$ such that

$$\begin{cases} \dot{k} = (\partial H / \partial q) \\ \dot{q} = -(\partial H / \partial k). \end{cases}$$

If we consider that $(d/dt)H(k, q) = (\partial H/\partial k)\dot{k} + (\partial H/\partial q)\dot{q} = 0$, we can deduce that $H(k, q)$ is constant under the flow. Hamiltonian systems are therefore a special class of conservative systems.

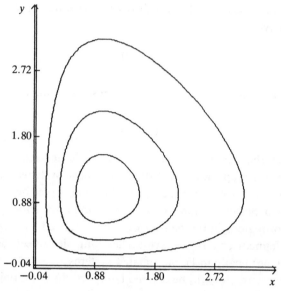

Fig. 2.2. Closed orbits generated by the LVG model.

interpretation, x denotes the level of employment (relative to population) and y denotes the share of wages in the national income. Thus the model can be used to investigate the dynamics of production and the distribution of income. By simple manipulations, one can verify that, along the trajectories of the system,

$$F(x, y) = e^{-dx}e^{-by}x^c y^a = \text{const.},$$

and conclude that the system is indeed conservative, even though no direct economic interpretation of F has been provided.

It can also be shown that the LVG system has two non-negative fixed points: the origin of the co-ordinate axes, which is a *saddle point*, and a second fixed point located in the positive quadrant, which is a *centre*.[10] All the trajectories starting in the positive quadrant are closed orbits around the fixed point. There are of course no attracting limit cycles, and initial conditions entirely determine which of the infinite number of cycles is actually followed. Any two-dimensional set of initial conditions is mapped by the action of the flow to another set of equal volume, although not necessarily of the same shape.

---

[10] A definition of saddle point and centre will be provided later in this chapter.

A second interesting economic example of a quasi-conservative (in fact quasi-Hamiltonian) system is given by the well-known model of infinite horizon optimal growth, which can be formulated as follows.

Consider the problem:

$$\max_{k} \int_0^{\infty} \mathcal{U}(k, \dot{k}) e^{-\rho t} dt,$$

$$(k, \dot{k}) \in S \subset \mathbb{R}^{2n}, \tag{2.6}$$

$$k(0) = k_0,$$

where: k denotes the capital stock; $\dot{k}$, net investment; the set S is convex and embodies the technological restrictions; $\rho \in \mathbb{R}^+$ is the positive discount rate; $\mathcal{U}$ is a concave utility function.

To attack this problem by means of the Pontryagin maximal principle, we must first of all introduce an auxiliary vector-valued variable $q \in \mathbb{R}^n$ and define a function[11]

$$H(k, q) \equiv \max_{k;(k,\dot{k})\in S} \{\mathcal{U}(k, \dot{k}) + q\dot{k}\}.$$

The necessary (though not sufficient) condition for problem (2.6) to be solved is that the time evolution of k and q satisfies the following system of differential equations:

$$\dot{k} = (\partial H / \partial q),$$

$$\dot{q} = -(\partial H / \partial k) + \rho q. \tag{2.7}$$

It will be noticed that system (2.7) can be thought of as a Hamiltonian system, plus a (linear) perturbation given by the term ρq. This has allowed economists to make use of a number of useful results on Hamiltonian systems. It must be observed, however, that in physics, the Hamiltonian function $H(q, k)$ is usually convex in both sets of variables, whereas in economics H is usually assumed to be concave in k and convex in q, which makes the economic problem radically different mathematically from the physical one. In particular, whereas stationary points in the latter problem are of the centre type, in economics, for the unperturbed Hamiltonian case ($\rho = 0$), the stationary points are all saddles.

Unlike conservative ones, *dissipative* dynamical systems, on which most of this book concentrates, are characterized by contraction of phase space volumes with increasing time. Dissipation can be formally described by the property that, for a system such as (2.5), the Lie derivative is negative,

---

[11] Notice that the Hamiltonian function $H(k, q)$ can be interpreted as the (maximum) current value of national income evaluated in terms of utility.

i.e., we have

$$\sum_{i=1}^{n} \frac{\partial \dot{x}_i}{\partial x_i} < 0,$$

or, equivalently

$$\sum_{i=1}^{n} \frac{\partial f_i(x)}{\partial x_i} < 0.$$

Because of dissipation, the dynamics of a system whose phase space is n-dimensional, will eventually be confined to a subset of Hausdorff dimension[12] smaller than n. Thus, in sharp contrast to the situation encountered in conservative systems, dissipation permits one to distinguish between *transient* and *permanent* behaviour.[13] For dissipative systems, the latter may be quite simple even when the number of phase space variables is very large.

To better understand this point, think of an n-dimensional system of differential equations characterized by a unique, globally asymptotically stable equilibrium point. Clearly, for such a system, the flow will contract any n-dimensional set of initial conditions to a zero-dimensional final state. Think also of an n-dimensional ($n \geq 2$) dissipative system characterized by a unique, globally stable limit cycle. Here, too, once the transients have disappeared, we are left with only one trajectory: the cycle, whose dimension is, of course, one.

The asymptotic regime of a dissipative system is the only observable behaviour, in the sense that it is not ephemeral, can be repeated and therefore be 'seen' (e.g., on the screen of a computer). The structure of the limit set of a dynamical system is often easier to investigate than its overall orbit structure. For most practical problems, moreover, an understanding of limit sets and their basins of attraction is quite sufficient to answer all the relevant questions.

Even though transients may sometimes last for a very long time and their behaviour may be an interesting subject for investigation, we shall deal with them only occasionally, and concentrate instead on the long-term behaviour of dynamical systems.

---

[12] The concept of Hausdorff dimension will be discussed below in Chapter 7.

[13] The effect of contraction of areas can be also investigated from the point of view of information theory. Broadly speaking, we may say that dissipation implies the loss of memory of initial conditions, so that they play no role in the dynamics of the system once this has reached its asymptotic regime. On this and related points, see Shaw (1981).

2.3 INVARIANT, α- AND ω-LIMIT, ATTRACTING SETS. ATTRACTORS

To discuss the asymptotic behaviour of dynamical systems, we need to develop some basic concepts which are common to both flows and maps and which in a sense describe different kinds or degrees of recurrence of the orbits of the system. We first define *invariant sets*.

Definition 2.1. *For a flow ϕ_t [or a map G] defined on $U \subset \mathbb{R}^n$, a subset $S \subset U$ is said to be invariant if*

$$\phi_t(S) \subset S \quad \forall\, t \in \mathbb{R}$$
$$[\text{resp.,} \quad G^n(S) \subset S \quad \forall\, n \in \mathbb{Z}].$$

Certain invariant sets play an important role in the organization of system orbits in the phase space. To fix ideas, consider a vector field

$$\dot{x} = f(x), \qquad x \in \mathbb{R}^n. \tag{2.8}$$

Let (2.8) have an equilibrium $\{\bar{x} \mid f(\bar{x}) = 0\}$ and let us consider the associated linear system

$$\dot{y} = Ay, \tag{2.9}$$

where $A \equiv D_x f(\bar{x})$ is a constant $n \times n$ matrix.

Suppose now that A has $s + u + c = n$ distinct eigenvalues with, respectively, negative, positive and zero real parts. Then $\mathbb{R}^n$ can be represented as the direct sum of three subspaces E^s, E^u and E^c, defined as follows:

$$E^s = \text{span}\{e_1, \dots, e_s\}$$
$$E^u = \text{span}\{e_{s+1}, \dots, e_{s+u}\}$$
$$E^c = \text{span}\{e_{s+u+1}, \dots, e_{s+u+c}\},$$

to which there correspond, respectively, the (generalized) eigenvectors $\{e_1, \dots, e_s\}$, $\{e_{s+1}, \dots, e_{s+u}\}$, $\{e_{s+u+1}, \dots, e_{s+u+c}\}$. These are called, respectively, *stable, unstable* and *centre* (or *central*) *eigenspace*, and are invariant under the flow generated by (2.8). Moreover, a result called 'the invariant manifold theorem' has demonstrated that in the original nonlinear flow, these eigenspaces are distorted into *stable, unstable* and *centre manifolds*, denoted respectively by W^s, W^u and W^c,[14] which are tangent to the corresponding eigenspaces at the equilibrium point and are invariant under the flow. We shall see that the centre manifold plays a particularly

---

[14] An n-dimensional manifold $M \subset \mathbb{R}^N$ is a set for which each $x \in M$ has a neighbourhood U for which there is a diffeomorphism $\phi: U \to \mathbb{R}^n$ ($n \leq N$). Thus, roughly speaking, we can say that a manifold is a set that *locally* has the structure of Euclidean space.

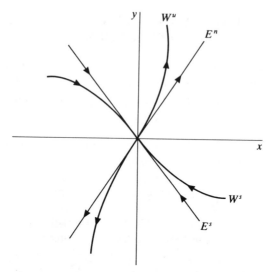

Fig. 2.3. Stable and unstable eigenspaces and manifolds in a planar dynamical system.

crucial role in the study of bifurcation. An analogous theorem exists for maps.

The stable and unstable manifolds are unique, the centre manifold need not be. Moreover, the direction of motion on the centre manifold cannot be determined from a purely linear argument.

Certain special configurations of stable and unstable manifolds, created by special orbits which are called *homoclinic* or *heteroclinic*, play a crucial role in the appearance of chaos in dynamical systems. A formal definition, which is common to flows and maps is the following:

Definition 2.2.   *Let S be an invariant set of a flow ϕ_t [a map G]. Let p be a point in the phase space of the flow [map] and suppose that $\phi_t(p) \to S$ as $t \to +\infty$ and as $t \to -\infty$ [resp., $G^n(p) \to S$ as $n \to +\infty$ and $n \to -\infty$]; then the orbit of p is said to be* homoclinic to S.

In the simple case in which S is a fixed point, a homoclinic orbit (of a flow) is depicted as in Fig. 2.4a.

Definition 2.3.   *Let S_1 and S_2 be two disjoint invariant sets of a flow ϕ_t [a map G] and suppose that $\phi_t(p) \to S_1$ as $t \to +\infty$ and $\phi_t(p) \to S_2$ as $t \to -\infty$ [resp., $G^n(p) \to S$, as $n \to +\infty$ and $G^n(p) \to S_2$ as $n \to -\infty$]; then the orbit of p is said to be* heteroclinic to S_1 and S_2.

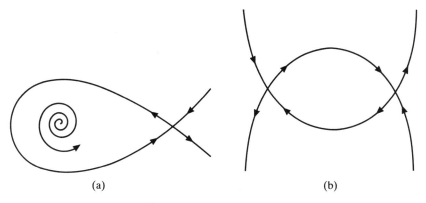

Fig. 2.4. (a) a homoclinic orbit; (b) a heteroclinic orbit.

Two heteroclinic orbits of a flow to two disjoint fixed points are depicted in Fig. 2.4b. If the invariant sets S, S_1, S_2 are hyperbolic, we can say that the orbit of p is homoclinic to S if p lies in both the stable and the unstable manifold of S, and that the orbit of p is heteroclinic to S_1 and S_2 if p lies in the stable manifold of S_1 and in the unstable manifold of S_2. In these cases p is called, respectively, a homoclinic or a heteroclinic point.[15]

A particularly interesting class of invariant sets includes those whose elements are asymptotic limit points. We can now provide the following:

Definition 2.4. *The ω-limit (α-limit) of x for a flow ϕ_t [resp. a map G] is the set of accumulation points of $\phi_t(x), t \to +\infty$ ($t \to -\infty$) [resp. the set of accumulation points of $G^n(x), n \to +\infty$, ($n \to -\infty$)].*[16]

Consider, for example, a system of o.d.e. characterized by two fixed

---

[15] A fixed (equilibrium) point x of a flow is said to be *hyperbolic* if the matrix of derivatives of the vector field at x has no eigenvalues with zero real parts. For a fixed point of a map, the requirement is that the matrix of the derivatives of the map at x has no eigenvalues with modulus equal to one. The idea can be generalized to points of invariant sets of a flow (or a map) which are not fixed points. If we consider a flow on a certain invariant set Λ containing no equilibrium points, hyperbolicity essentially entails that, for each point $x \in \Lambda$, apart from the direction of the flow itself, the transverse directions can be split into those along which the flow expands away from trajectories in Λ and those along which it contracts towards trajectories in Λ. An analogous consideration can be made for maps. A more rigorous and complete definition can be found, for example, in Guckenheimer and Holmes (1983, p. 238) and Wiggins (1990, pp. 463–70). A moment's reflection will suggest that there is no reason to expect that, in general, flows or maps investigated in applications should have hyperbolic structures.

[16] Notice that y is an accumulation point of $\phi_t(x), t \to \infty$, if there is a sequence $t_i \to \infty$ such that $\phi_{t_i}(x) \to y$ (analogously for a map).

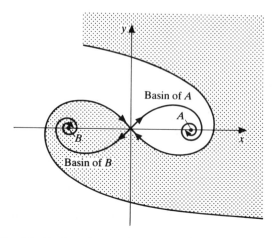

Fig. 2.5. Basins of attraction of two stable fixed points.

points, $\bar{x}_1$ (stable) and $\bar{x}_2$ (unstable), and no other limit set. Then we shall say that for such a system $\bar{x}_1$ is the ω-limit and $\bar{x}_2$ the α-limit of the flow.[17]

Not all ω-limit sets of the system are equally interesting, and in most cases of practical interest we wish to locate a region of the phase space which ultimately captures all the orbits originating in a certain (not too small) domain. This idea can be made more precise by the following:

Definition 2.6.　　*A compact set $A \subset U$ is said to be an* attracting set *if:*

(i)　*A is invariant under ϕ_t [resp., G];*

(ii)　*A has a shrinking neighbourhood, i.e., there exists an open neighbourhood V of A, s.t. for all $x \in V$, $\phi_t(x) \in V$ [resp., $G^n(x) \in V$] for all $t \geq 0$ [resp., $n \geq 0$] and $\phi_t(x) \rightarrow A$ [resp., $G^n(x) \rightarrow A$] for $t \rightarrow +\infty$ [resp., $n \rightarrow +\infty$].*

---

[17] Weaker forms of recurrence are sometimes used in applications and we need the following:

Definition 2.5.　　*A point p is said to be* non-wandering *for a flow ϕ_t [a map G], if for any open neighbourhood U of p and $T > 0$ [resp., $N > 0$] there exists a $t > T$ [resp., $n > N$], such that $\phi_t(p) \in U$ [resp., $G^n(p) \in U$].*

The non-wandering set is the set of all non-wandering points. An example of a non-wandering set occurring for planar vector fields is given by a homoclinic orbit, i.e., an orbit forming a loop which contains a single saddle point (see Fig. 2.4a). Each point of the orbit has neighbourhoods that always come back near the point.

For an attracting set we define a *basin of attraction* as

$$D = \bigcup_{t \leq 0} \phi_t(V), \qquad [\text{resp.,} \bigcup_{n \leq 0} G^n(V)].$$

D is also called the *stable manifold* of *A*. We can define analogously a *repelling set* and its *unstable manifold* by replacing *t* with −*t* and *n* with −*n*. The study of repelling sets (and the associated *repellors*) can provide very useful information on the dynamics of a system. Notice, however, that 'non-attracting' is *not* the same thing as 'repelling'. Saddle-type invariant sets are neither attractors nor repellors, and with regard to them it is possible to define stable *and* unstable manifolds.

The fact that a set is attracting does not mean that all its parts are attracting too. Therefore, in order to describe the asymptotic regime of a system we need the stronger concept of *attractor*. Strangely enough, there is no straightforward and universally adopted definition of attractors, and although the properties of the simpler cases can be easily dealt with, more complicated types of attractors present difficult conceptual problems.

In a non-rigorous, operational sense, an attractor is a set on which experimental points generated by a flow or a map accumulate for large *t*.[18] In the absence of a universal consensus on a mathematical definition of attractor, we shall adopt here a 'naive' definition as follows:[19]

Definition 2.7.   *A set A is said to be an* attractor *for a flow ϕ_t [resp., a map G] if it enjoys the properties of an attracting set, according to Definition 2.6, and in addition it is topologically transitive.*

A compact invariant set *A* is said to be *topologically transitive* for a flow *φ* [resp., for a map *G*] if, for any two open sets $U, V \subset A$, $\phi_t(U) \cap V \neq \emptyset$, for all $t \in \mathbb{R}$ [respectively, $G^n(U) \cap V \neq \emptyset$, for all $n \in \mathbb{Z}$]. Topological transitivity implies (and is implied by) the existence of a *dense orbit*[20] on *A* for the flow ϕ_t [resp., the map *G*].

The presence of topological transitivity (or, equivalently, a dense orbit)

---

[18]   Cf. Eckmann and Ruelle (1985, p. 623).

[19]   This definition corresponds essentially to those given, for example, by Eckmann (1981, p. 644) and Wiggins (1990, p. 45). The term 'naive' is used by Guckenheimer and Holmes (1983), to whose work we refer the reader for a discussion of some of the problems arising with the definition above, and for a subtler definition, involving probabilistic elements (Guckenheimer and Holmes, 1983, pp. 256–7). See also on this point, Eckmann and Ruelle (1985, pp. 623 and 645) and the works quoted there.

[20]   A subset *S* of a space *M* is said to be dense in *M* provided its closure is the entire space *M*. Intuitively, the presence of a dense orbit in a set implies that, under the action of a flow (or a map), the orbit moves from one arbitrarily small neighbourhood to any other and thus it tends to 'fill' the set.

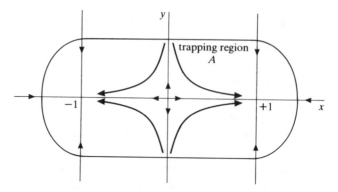

Fig. 2.6. Attracting sets and attractors.

implies that the set A is indecomposable in dynamical terms, i.e., it cannot be divided into two or more invariant sets.

Whether or not the requirement of indecomposability is relevant in applications cannot be decided *a priori*. A case often quoted in the literature[21] in which the former alternative is obviously true is illustrated in Fig. 2.6. The diagram is derived from the well-known system of o.d.e.:

$$\dot{x} = x - x^3, \quad x \in \mathbb{R},$$
$$\dot{y} = -y, \quad y \in \mathbb{R}.$$

The reader can promptly ascertain that, although the interval ($x \in [-1, 1]$, $y = 0$) is an attracting set, almost all points in the (x, y) plane will end up near one or the other of the two attractors, i.e., $(\pm 1, 0)$.

In other cases, however, it is not clear whether the attracting set contains a true attractor or a large number of stable subsets (typically periodic orbits) with basins of attraction so small and so convoluted that small perturbations (e.g., small computing errors) prevent one from actually observing the true periodic asymptotic behaviour of the system. In these situations, we may wonder whether the requirement of inde-composability and the distinction between attracting sets and attractors remain practically relevant.[22] An example is provided by the chaotic attractor numerically investigated in Chapter 12.

---

[21] See Guckenheimer and Holmes (1983, p. 257); Eckmann and Ruelle (1985, pp. 623 and 645).

[22] Notice here that there are authors who would call 'attractor' our 'attracting set', and 'minimal attractor' our 'attractor'. There are also authors who do not distinguish between attracting sets and attractors at all. See, for example, Temam (1988, pp. 20–1).

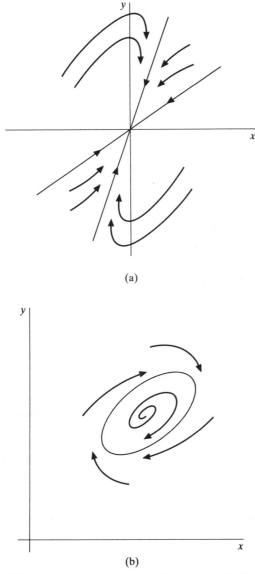

(a)

(b)

Fig. 2.7. (a) A stable fixed point; (b) a stable limit cycle.

Remark 2.1.   Attractors, as experimental objects, should also enjoy the further property of being stable under small random perturbations, like the round-off errors due to the floating-point truncation in computer

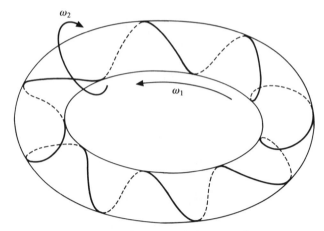

Fig. 2.8. A quasiperiodic orbit on a T^2 torus.

calculations, or the external noise in experimental simulations (see Ruelle, 1989, p. 25). In this sense, numerical evidence of the presence of attractors is very valuable.

2.4 BASIC TYPES OF ATTRACTORS

Much of our investigation will be devoted to the study of qualitative as well as quantitative properties of orbits on attractors (or attracting sets). These can be classified into a number of basic types which are important in applications:

(*a*) *Equilibrium points* (synonyms: fixed points, stationary points, rest points, singular points, critical points) (see Fig. 2.7a)

This is the simplest type of attractor which consists of points $\bar{x}$ in the phase space, such that for the differential equation $\dot{x} = f(x)$, we have:

$$0 = f(\bar{x}),$$

or, for the map $x \mapsto G(x)$, we have

$$\bar{x} = G(\bar{x}).$$

(*b*) *Periodic orbits* (limit cycles) (see Fig. 2.7b)

Periodic orbits represent the next level of complexity of asymptotic motion. For an ordinary differential equation, we say that its solution is periodic if, for some positive T, $\phi_T(x) = x$. The closed orbit $\gamma =$

$\{\phi_t(x)|t \in [0, T)\}$ is called a stable (unstable) *limit cycle*, if there exists a neighbourhood U of γ such that $\lim_{t \to +(-)\infty}$, for all $x \in U$. The smallest possible $T > 0$ for which this property holds is called the *period* of the orbit and

$$\frac{\omega}{2\pi} = \frac{1}{T}$$

is its *frequency*, measured as number of cycles per unit of time.[23] For maps, a period-n point $\bar{x}$ is a point such that $G^n(\bar{x}) = \bar{x}$. The orbit in this case is a sequence of n distinct points $\{\bar{x}, G(\bar{x}), \ldots, G^{n-1}(\bar{x})\}$ which, under the iterated action of G, are repeatedly visited by the system, always in the same order.

(c) Quasiperiodic orbits

Roughly speaking, we can think of quasiperiodic orbits as the output of two or more oscillators, the ratios of whose frequencies are irrational. Formally, we can write the following:

Definition 2.8. *A function* $h:\mathbb{R} \to \mathbb{R}^n$ *is called* quasiperiodic *if it can be written in the form* $h(t) = H(\omega_1 t, \ldots, \omega_n t)$, *where* H *is periodic of period* 2π *in each of its arguments, and two or more of the* n *frequencies are incommensurable (i.e., their ratios are irrational numbers).*

Thus we call quasiperiodic a continuous-time dynamical system, whose solution $\phi(t)$ is quasiperiodic in time. The orbits associated with an n-frequencies quasiperiodic (continuous-time) system are defined on a *torus*[24] of dimension n, embedded in an m-dimensional space ($m \geq n$), with $\omega_1 t, \omega_2 t, \ldots, \omega_n t$ serving as co-ordinates of the torus. A geometrical insight on the simplest case ($n = 2$) is provided by Fig. 2.8. Quasiperiodic orbits of *maps* can also be defined analogously.[25]

Examples of quasiperiodic dynamical systems in economics can be found, e.g., in H.W. Lorenz (1988b) and Puu (1987), as well as in the model investigated in Chapter 12.

(d) Chaotic or strange attractors

Loosely speaking, these are attractors with an orbit structure that is more complicated than that of periodic or quasiperiodic systems. The under-

---

[23] Naturally, ω is the frequency measured in radians per unit of time.

[24] A n-dimensional torus, denoted by T^n, is the product of n-circles ($n > 1$). Intuitively, a two-dimensional torus T^2 can be thought of as the surface of the inner tube of a car tyre, which is of course 'embedded' in the $\mathbb{R}^3$ ordinary space.

[25] Sometimes the term *almost periodic* function is used to denote quasiperiodic functions with an infinite number of basic frequencies.

standing, detection and measurement of chaotic or strange attractors are the object of most of the following chapters, so that we shall limit ourselves here to a few brief considerations. A chaotic attractor arises typically when the overall contraction of volumes, which characterizes dissipative dynamical systems, takes place by shrinking in some of the directions, accompanied by (less rapid) stretching in some of the others. This fact has profound consequences, as it implies that there may be an unstable motion even *within the attractor*.

An important consequence of this instability is that pairs of orbits which originate from points arbitrarily near one another on the attractor become exponentially separated as time goes by. Thus, arbitrarily small errors of measurement of initial conditions are magnified by the action of the flow (or the map), so that accurate prediction of the future course of the orbits becomes impossible, except in the short run.

The situation just described is referred to as *sensitive dependence on initial conditions* and can be defined formally as follows:

Definition 2.9.   *A flow* $\phi_t: U \rightarrow U$ *[or a map* $G: U \rightarrow U$*] where* $U \subseteq \mathbb{R}^n$, *has* sensitive dependence on initial conditions (SDIC) *if there exists* $\delta > 0$ *such that, for any* $x \in U$ *and any neighbourhood* N *of* x, *there exists* $y \in N$ *and* $t \geq 0$ *[resp.,* $n \geq 0$*] such that* $|\phi_t(x) - \phi_t(y)| > \delta$ *[resp.,* $|G^n(x) - G^n(y)| > \delta$*]*.

Hence we shall write:

Definition 2.10.   *An attractor, as described in Definition 2.7, is called* chaotic *or* strange *if it has* SDIC.

At this time we should like to add two more remarks.

Remark 2.2.   There is no universally adopted definition of strange attractors. In particular, there is no consensus on the use of the terms 'strange' and 'chaotic'. Some authors (e.g., Grebogi *et al.*, 1984) use the former concept to refer to the geometry of the attractor (i.e., to its being a fractal object) and the latter to refer to its dynamics. Others (e.g., Eckmann and Ruelle, 1985, p. 625) disagree and define 'strange' (or equivalently, chaotic) an attractor characterized by SDIC, on the ground that the dynamics of the attractor are a more important aspect than its geometry. We shall not take up this controversy here, but we shall generally follow Eckmann and Ruelle's usage. Also, the definition some authors use of strange (or chaotic) attractors includes the requirement that periodic orbits are dense on the attractor (see Devaney (1986, p. 50)). For a definition of strange attractors close to ours, but with

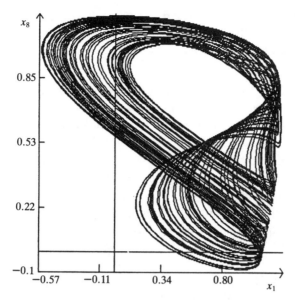

Fig. 2.9. A chaotic attractor.

emphasis on the role played by homoclinic orbits in generating chaos, see Guckenheimer and Holmes (1983, p. 256).

Remark 2.3. Although our main object here is to study chaotic attractors, the investigation of chaotic invariant sets which are not attracting can be extremely useful in understanding the origin and structure of strange behaviour. Besides, the presence of such invariant sets can dramatically affect the overall behaviour of the system concerned. This remark will be recalled when we discuss the so-called horseshoes and homoclinic orbits (see Chapter 8).

2.5 STABILITY

The study of attracting sets is intimately linked with that of stability of orbits. Given the vastness of the subject, we shall provide here only a few definitions and a brief discussion of two basic methods of determining stability which are used in this book (as well as in the economic literature).

Consider the system of o.d.e.

$$\dot{x} = f(x), \qquad x \in \mathbb{R}^n, \tag{2.10}$$

where $f: U \to \mathbb{R}^n$ with U an open subset of $\mathbb{R}^n$ and f is $C^r, r \geq 1$.

Definition 2.11.   *A solution $\phi(t)$ of (2.10) is said to be (Lyapunov) stable if, given $\epsilon > 0$, there exists a $\delta(\epsilon) > 0$ such that, for any other solution $\psi(t)$ of (2.10) for which $|\phi(t_0)-\psi(t_0)| < \delta$, we have $|\phi(t)-\psi(t)| < \epsilon$ for all $t \geq 0$.*

Definition 2.12.   *Assume that an orbit $\phi(t)$ is stable according to the definition above and, moreover, that there exists $\bar{\delta} > 0$ such that if $|\phi(t_0) - \psi(t_0)| < \bar{\delta}$, then $\lim_{t \to +\infty} |\phi(t) - \psi(t)| = 0$. Then $\phi(t)$ is said to be* asymptotically stable.[26]

In simple terms, this means that, if a stable orbit is slightly perturbed, the ensuing motion will not diverge much from that of the unperturbed orbit. If, in addition, the orbit is asymptotically stable, the effect of perturbation will be progressively eliminated as time goes by. Analogous definitions can be written for maps, thus:

Definition 2.13.   *A solution $\{x\}_{i=0}^n$ of a map $x \mapsto G(x)$, where $\{x\}_{i=0}^n \equiv \{x_0, x_1 \equiv G(x_0), x_2 \equiv G^2(x_0), \ldots, x_n \equiv G^n(x_0)\}$ is (Lyapunov) stable if, for any given $\epsilon > 0$, there exists a $\delta(\epsilon) > 0$ such that, for any other solution $\{y\}_{i=0}^n$ with $|x_0 - y_0| < \delta$, we have $|G^n(x_0) - G^n(y_0)| < \epsilon$ for $n \in [0, +\infty)$.*

Definition 2.14.   *The solution $\{x\}_{i=0}^n$ is asymptotically stable if it is stable and, in addition, there exists a $\bar{\delta} > 0$ such that if $|x_0 - y_0| < \bar{\delta}$, $\lim_{n \to \infty} |G^n(x_0) - G^n(y_0)| = 0$.*[27]

2.5.1 Linearization

In applications, two basic methods are used to ascertain stability. The first of these is referred to as *linearization*.

Consider again a solution $\phi(t)$ of system (2.10) and introduce a new, vector-valued variable

$$y \equiv x - \phi(t).$$

We can write then:

$$\dot{x} = f[y + \phi(t)]. \tag{2.11}$$

If we perform an expansion in a Taylor series of $f(y + \phi(t))$ about $y = 0$

---

[26] Notice that, by reversing the motion in time, asymptotically stable orbits become unstable. The reverse, however, is not necessarily true, since instability of the saddle type remains such for both forward and backward motion. This remark becomes relevant, for example, when we try to locate an unstable fixed point on the screen of our computer. This operation is quite possible when the fixed point is of a node or a focus type: all we have to do is run our program in reverse. But we shall fail if the point is a saddle. An analogous argument can be used with regard to limit cycles.

[27] If the map G is invertible, unstable fixed points which are not of the saddle type may be located by iterating the map in reverse.

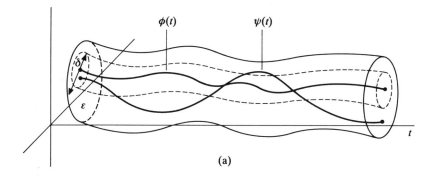

(a)

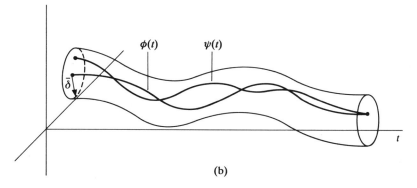

(b)

Fig. 2.10. (a) Lyapunov stability; (b) asymptotic stability.

and neglect all nonlinear terms, we obtain the (time-dependent) linear equation

$$\dot{y} = Df[\phi(t)]y. \tag{2.12}$$

Although there is no general method for solving time-dependent linear differential equations, thère are some basic results for the case in which the time dependence of $\phi(t)$ is particularly simple. For example, if $\phi(t) = \bar{x} = $ constant, equation (2.12) can be written as

$$\dot{y} = Ay, \tag{2.13}$$

where $A \equiv Df(\bar{x})$ is a constant matrix. In this case, the general solution of (2.13) can be written explicitly as

$$y(y_0, t) = e^{tA}y_0, \tag{2.14}$$

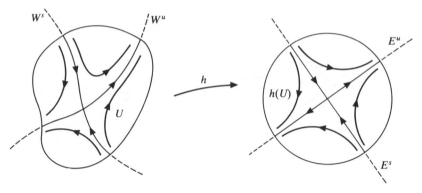

Fig. 2.11. The Hartman–Grobman theorem.

where e^{tA} is a constant matrix defined by the convergent series

$$[I + tA + \frac{t^2}{2!}A^2 + \ldots + \frac{t^n}{n!}A^n + \ldots].$$

Then we can apply the following theorem:[28]

Theorem 2.1 (Hartman–Grobman).   *If A has no zero or purely imaginary eigenvalues, then there is a homeomorphism h defined on some neighbourhood U of $\bar{x}$ in $\mathbb{R}^n$, locally taking orbits of the nonlinear flow ϕ_t of (2.10) to those of the linear flow (2.14). The homeomorphism preserves the sense of orbits and can also be chosen to preserve parametrization by time.*

Theorem 2.1 can be heuristically interpreted by saying that – when the prescribed conditions apply – the local behaviour of a nonlinear system is qualitatively similar to that of the linearized one. In particular, Theorem 2.1 implies that asymptotic stability (or instability) of (2.13) corresponds to local asymptotic stability (or instability) of (2.10).

It can be easily shown that a necessary and sufficient condition for the linearized system to be asymptotically stable is that all the eigenvalues of the matrix A have negative real parts.

The orbit structure of the linearized system (2.13) (and therefore, whenever Theorem 2.1 holds, the *local* orbit structure of (2.10)) depends on the eigenvalues of the matrix A and, at least for n small, a geometrical representation can be very illustrative of the behaviour of the system.

In the simple case in which $n = 2$, and A has two distinct eigenvalues, respectively λ_1 and λ_2, the following cases may be identified:

[28] See Hartman (1964); also Guckenheimer and Holmes (1983, p. 13).

(1) λ_1 and λ_2 real.

   (1.1) $\lambda_1 + \lambda_2 < 0$, $\lambda_1 \lambda_2 > 0$. The equilibrium point $y = 0$ is stable and the orbits approach it from a definite direction. The equilibrium in this case is called a *stable node* (Fig. 2.12a).

   (1.2) $\lambda_1 + \lambda_2 > 0$, $\lambda_1 \lambda_2 > 0$. This is the same as 1.1, when we reverse the motion in time. The equilibrium is called an *unstable node* (Fig. 2.12b).

   (1.3) $\lambda_1 \lambda_2 < 0$. All the orbits eventually move away from equilibrium, except those originating in points located on the stable branches. The equilibrium is called a *saddle point* (Fig. 2.12e).

(2) λ_1 and λ_2 form a pair of conjugate complex numbers (i.e., $\lambda_1 = \gamma + i\omega$; $\lambda_2 = \gamma - i\omega$, γ and ω being real numbers).

   (2.1) $\gamma < 0$. The equilibrium point is approached in an oscillatory manner for $t \to \infty$ and is called a *stable focus* (Fig. 2.12c).

   (2.2) $\gamma > 0$. The equilibrium point is approached in an oscillatory manner for $t \to -\infty$ and is called an *unstable focus* (Fig. 2.12d).

An interesting special subcase of (2) is obtained when $\gamma = 0$. In this case, *for the linear system* (2.13), the orbits form a continuum of close curves surrounding the equilibrium point, which is called a *centre* (Fig. 2.12f). However, in this case, Theorem 2.1 does not hold and system (2.13) does not provide sufficient information concerning the local behaviour of the original nonlinear system (2.10). The time behaviour of the latter must be ascertained by considering higher order terms, and it may be convergence to equilibrium (the so-called *weak focus*) or divergence from it.

If $n = 3$, more complicated situations may arise which we shall not describe in any detail here. A particularly interesting case is given by the combination of nodes, or foci, and saddles, which gives rise to the so-called saddle-nodes or saddle-foci which are depicted in Fig. 2.13. A case of saddle-focus will appear again later, in Chapter 8, when we discuss the so-called Šilnikov phenomenon, as well as in the applications investigated in Chapters 11 and 13.

A discrete-time version of the Hartman–Grobman theorem exists[29] which can be used to establish the stability of periodic orbits of a map

$$x \mapsto G(x), \qquad x \in \mathbb{R}^n, \tag{2.15}$$

[29] See Hartman (1964).

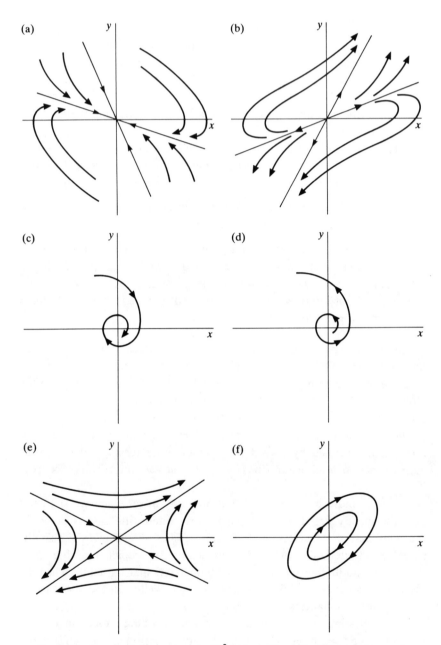

Fig. 2.12. Local orbit structure in $\mathbb{R}^2$: (a) stable node; (b) unstable node; (c) stable focus; (d) unstable focus; (e) saddle point; (f) centre.

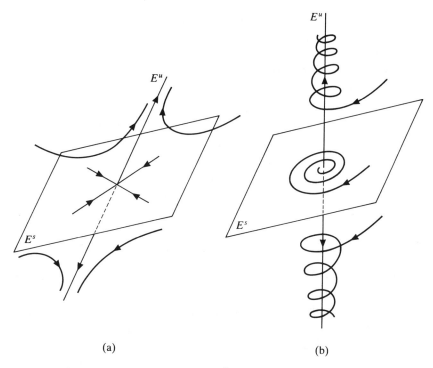

Fig. 2.13. Local orbit structure in $\mathbb{R}^3$: (a) saddle-node; (b) saddle-focus.

defined on some open subset $U \subset \mathbb{R}^n$. A point $p \in U$ is a period-n point of G if:

$$G^n(p) \equiv \underbrace{G(G\ldots(G(p)))}_{n \text{ times}} = p.$$

Consider now that, in this case, $p_1 \equiv G(p)$, $p_2 \equiv G^2(p)$, ..., $p_n \equiv G^n(p)$ are each fixed points for the map:

$$x \mapsto G^n(x). \tag{2.16}$$

By means of an argument similar to that followed with regard to o.d.e., we can associate with (2.16) the following linear map:

$$y \mapsto By, \qquad y \in \mathbb{R}^n, \tag{2.17}$$

where $B = DG^n(p_j)$, for any $j = 1,\ldots,n$; $DG^n(p_j) = DG(G^{n-1}(p_j))$ $DG(G^{n-2}(p_j))\ldots DG(p_j)$ (by the chain rule), and $y = 0$ is a fixed point of the map. We can then write:

Theorem 2.2.   *If p_0 is a period-n point for map (2.15), i.e., we have $\{p\}_{i=0}^{n} = \{p_0, p_1, \ldots, p_n = p_0\}$, and the matrix $DG^n(p)$ has no eigenvalues of modulus one, then asymptotic stability (instability) of the fixed point $y = 0$ of (2.17) corresponds to local asymptotic stability (instability) of the periodic orbit $\{p\}_{i=0}^{n}$.*

Notice that in the case of maps, the question of stability (instability) of a periodic orbit is reduced to that of stability (instability) of a fixed point. Stability of the linear map (2.17) can be ascertained by verifying that all eigenvalues of matrix B have modulus less than one.

The application of the linearization technique to the case in which $\phi(t)$ is periodic in time will be discussed in Chapter 4, when we deal with Poincaré maps.

2.5.2 Lyapunov direct method

The second method for determining stability is the so-called second (or direct) method of Lyapunov which can be briefly described thus.

Consider again the system of o.d.e. (2.10). Suppose (2.10) has a fixed point, which, without loss of generality, we can locate at $x = 0$. Then we can write the following:

Theorem 2.3.   *If there exists a C^1 function $V(x): W \to \mathbb{R}$, defined on some neighbourhood W of 0, such that*

(i) $V(0) = 0,$
(ii) $V(x) > 0$ for $x \neq 0,$
(iii) $\dot{V}(x) \leq 0$ in $W - \{0\},$

then $x = 0$ is stable. Moreover if

(iv) $\dot{V}(x) < 0$ in $W - \{0\},$

then $x = 0$ is asymptotically stable.

The derivative of $V(x)$ w.r.t. time must be taken along solution curves of (2.10), i.e., we have:

$$\dot{V}(x) = \sum_{i=1}^{n} \frac{\partial V}{\partial x_i} \dot{x}_i = \sum_{i=1}^{n} \frac{\partial V}{\partial x_i} f_i(x).$$

The Lyapunov 'second method' can also be extended to discrete-time systems (maps), as follows. Let us consider the map

$$x \mapsto G(x), \qquad x \in \mathbb{R}^n. \tag{2.18}$$

Suppose that there exists a fixed point $\bar{x}$ of G such that $\bar{x} = G(\bar{x})$.

Suppose also that we have a C^1 function $V(x)$, $V: W \to \mathbb{R}$ defined on some neighbourhood W of $\bar{x}$, such that

(i) $\quad V(\bar{x}) = 0$,

(ii) $\quad V(x) > 0$, for $x \neq \bar{x}$,

(iii) $\quad V(G(x)) - V(x) \leq 0$, in $W - \{\bar{x}\}$.

Then $\bar{x}$ is a stable fixed point of (2.18). Furthermore, if strict inequality holds in (iii) above, then $\bar{x}$ is asymptotically stable.

Remark 2.4. Stability established by means of the Lyapunov direct method is not local since the basin of attraction of 0 (or $\bar{x}$) contains the entire set W. Moreover, if we can put $W = U = \mathbb{R}^n$, then $x = 0$ (or $\bar{x}$) is said to be globally asymptotically stable.

Remark 2.5. Lyapunov functions are not unique. Although it can be proved that for stable systems some Lyapunov function does exist, there are no general methods available for finding a suitable one, and much is left to the investigator's skill and experience.

Remark 2.6. Lyapunov functions can also be used to define the so-called *trapping regions*. For vector fields, a trapping region is a closed connected region A of the phase space such that the motion of the system on the boundary of A is directed everywhere inward, and consequently all the nearby orbits are captured. An analogous definition can be given for maps. In applications, finding a trapping region is often the first step to locate attracting sets.

2.6 DYNAMICALLY EQUIVALENT SYSTEMS. CONJUGACY

When investigating a dynamical system S, we often find that, although it may impossible to provide a satisfactory analysis of S directly, a certain relation of equivalence can be established between S and another auxiliary system T which, on the contrary, is well understood. In a sense, this is what we have just done when we have discussed local stability and made recourse to the Hartman–Grobman theorem. In this section we would like to make the concept of equivalence a little more precise and to introduce a few concepts and definitions which will be used in various places in the following chapters.

Definition 2.15. *Let f and g be two C^r maps. Then f and g are said to be C^k-conjugate (or to have* differential equivalence*) if there exists a C^k diffeomorphism $(k \leq r)$ $h: \mathbb{R}^n \to \mathbb{R}^n$ such that $g \circ h = h \circ f$. If h is not* *defined on all* $\mathbb{R}^n$, *but only locally (e.g., about a given point)* f *and* g *are said to be* locally conjugate.

This situation is often described by the following diagram:

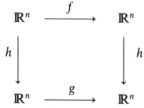

That is to say, when f and g are conjugate, orbits of f map to orbits of g under h and consequently their dynamics are equivalent.

Analogous (but not identical) results exist for flows, as indicated by the following:

Definition 2.16.   *Let* f *and* g *be two* C^r *vector fields on* $\mathbb{R}^n$. *Then* f *and* g *are said to be* C^k-*equivalent* $(k \leq r)$ *if there exists a* C^k *diffeomorphism* h *which takes orbits* $\phi(t, x)$ *generated by* f *to orbits* $\psi(t, x)$ *generated by* g, *preserving the orientation of the flow.*

Here, too, the dynamics of the two equivalent vector fields are essentially the same. In particular, when f and g are C^k-equivalent we have the following results:

(i)   if x_0 is a fixed point of f, then $h(x_0)$ is a fixed point of g;

(ii)   the eigenvalues of the matrix $Df(x_0)$ are the same (but for a positive multiplicative constant) as those of $Dg(h(x_0))$;[30]

(iii)   periodic orbits of f map to periodic orbits of g;[31]

(iv)   the image under h of the ω-limit set of an orbit $\phi(t, x)$ of f is the ω-limit set of the orbit $\psi(t, x)$ of g, and similarly for the α-limit set.

For both maps and flows, when $k = 0$, a more restricted form of equivalence obtains, called C^0-*conjugacy* (or *topological conjugacy* or *topological equivalence*). In this case, certain relevant topological facts, for example the distinction between nodes and foci, are lost under the h transformation.

The results just discussed provide powerful instruments to investigate apparently intractable problems. Suppose we want to study the behaviour of the map (or the vector field) f. Instead of attacking the problem directly, which may prove impossible, we can instead try to define an

---

[30] They are exactly the same if h also preserves parameterization by time. In this case f and g are called C^k-*conjugate.*

[31] The periodic orbits will have the same period if f and g are C^k-conjugate.

equivalence relation (a diffeomorphism *h*) between *f* and another map (or vector field) *g*, whose properties we know already, or we are able to ascertain. Since the dynamics of *f* and *g* are equivalent, in this way an apparently insurmountable obstacle may be overcome. As we shall see in Chapter 8 this idea is at the origin of the application of *symbolic dynamics* to the investigation of dynamical systems.

The concepts of conjugacy and equivalence are related to that of *structural stability* and *transversality* which have played an important role in the development of dynamical system theory. Before leaving this section, we want to provide some definitions and introductory ideas.

Whereas the term stability (or dynamic stability) refers to perturbations in the phase space, structural stability refers to perturbations in the function space. Broadly speaking, a dynamical system (a flow or a map) is said to be structurally stable if its dynamics are equivalent to that of a system sufficiently close to it in some sense.

At first sight, this seems to be a desirable property since we can never be certain that our mathematical idealizations correspond exactly to the real systems we are investigating, and we do not want our results to depend crucially on the exact functional form of our equations. However, we shall see in a moment, in our discussion of bifurcation, that there exists a different, more insightful point of view of this problem.

A formal definition of structural stability requires some preliminary clarification of the concept of *closeness* of dynamical systems. Let *M* be an *n*-dimensional, compact manifold; let $C^r(M, M)$ denote the space of C^r maps of *M* into *M*, so that elements of C^r can be thought of as vector fields; let us denote the subset of $C^r(M, M)$ consisting of C^r diffeomorphisms by $\text{Diff}^r(M, M)$. Then, two elements of $C^r(M, M)$ are said to be C^k-close if, along with their *k* derivatives, they are within a distance ϵ as measured in some norm. Then we can write the following:

Definition 2.17. *A map $G \in \text{Diff}^r(M, M)$ [or a vector field f in $C^r(M, M)$] is said to be* structurally stable *if there exists an ϵ-neighbourhood N of G [resp., of f] such that G is C^0-conjugate [resp., C^0-equivalent] to every map [resp., every vector field] in N.*

From the point of view of applications, it would be nice to know that structural stability is a generic property of dynamical systems and to determine rigorous criteria to ascertain its presence in specific cases. Unfortunately, neither of these goals has been attained. We cannot discuss here the problems encountered by mathematicians in the attempt to clarify this matter.[32] We shall limit ourselves to a few concise remarks.

Remark 2.7.   Precise conditions for structural stability have been established (Peixoto, 1962) for two-dimensional vector fields.

Remark 2.8.   For higher dimensional systems our understanding of the situation concerning structural stability and bifurcations varies inversely with the complexity of the system. We know a lot of systems whose non-wandering set contains only equilibrium points or periodic orbits. In these cases, structural stability may be related to certain local properties, i.e., to the fact that the eigenvalues of certain matrices do not have vanishing real parts (or that they are not equal to one in modulus). For systems having complex global motions, such as homoclinic intersections and chaotic invariant sets, there are no general criteria to verify whether a specific system is structurally stable (or, in some sense, generic).

The concept of transversality deals with intersections of surfaces or manifolds and, like that of structural stability, it has been employed to define generic properties of dynamical systems. Transverse intersections are generic events in the sense that they retain their topology under perturbations. Two or more manifolds which are transversal are also said to be in *general position*, which suggests that this situation is, in some sense, a likely, non-rare event.

Formally, we can write the following

Definition 2.18.   *Let x be a point in $\mathbb{R}^n$, M and N two differentiable manifolds in $\mathbb{R}^n$, and $T_x M, T_x N$ the tangent spaces of M and N, respectively, at x. Then M and N are* transversal *at x if $x \notin M \cap N$, or, if $x \in M \cap N$, then $T_x M + T_x N = \mathbb{R}^n$. M and N are* transversal *if they are transversal at every point $x \in \mathbb{R}^n$.*

The presence of this property can be verified by means of the concept of *codimension*. In general, the codimension of an l-dimensional submanifold of an n-dimensional manifold is $(n - l)$. Thus, two manifolds M and N in $\mathbb{R}^n$, intersecting at x will be transversal if

$$\text{cod}(T_x M) + \text{cod}(T_x N) = \text{cod}(T_x M \cap T_x N).$$

To see this in a simple case, consider the intersection of a two-dimensional surface (S_1) and a curve (S_2) in $\mathbb{R}^3$. In the case illustrated by Fig. 2.14a,

---

[32] The interested reader can consult Guckenheimer and Holmes (1983, pp. 38–42); also Wiggins (1990, pp. 56–62) on which our discussion is based, as well as the bibliography quoted there.

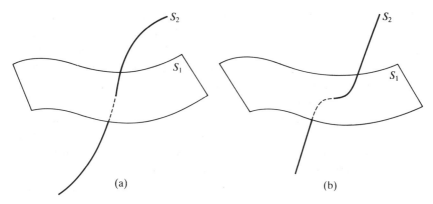

Fig. 2.14. (a) Transversal and (b) non-transversal intersections.

the curve is 'piercing' the surface, and we have

$$\mathrm{cod}(T_xS_1) + \mathrm{cod}(T_xS_2) = \mathrm{cod}(T_xS_1 \cap T_xS_2) = 3$$

Therefore the intersection is transversal.

On the contrary, in the case illustrated by Fig. 2.14b, the curve S_2 is tangential to S_1 at the point of intersection, and T_xS_2 is a line lying in T_xS_1. Therefore we shall have

$$\mathrm{cod}(T_xS_1) + \mathrm{cod}(T_xS_2) \neq \mathrm{cod}(T_xS_1 \cap T_xS_2) = 2,$$

and consequently the intersection will not be transversal.

It is clear that the non-transversal intersection is not generic, as it can be transformed into a transversal one by means of an arbitrarily small perturbation, but the reverse is not true. Small perturbations of the surfaces intersecting transversally will not destroy this property.

2.7 BIFURCATIONS

Bifurcation theory is a very complex research programme, which was originally outlined by Poincaré and which, in spite of very significant progress in the recent past, is far from complete. In this book, we shall only introduce the basic concepts with a view to applications, leaving aside the deeper theoretical questions.[33] In the preceding section, we have seen that, by definition, small perturbations of structurally stable systems do not yield qualitatively new dynamical phenomena. On the

---

[33] A more advanced and detailed treatment of the matter can be found, for example, in Wiggins (1988) and (1990, ch. 3), from which we have drawn heavily.

other hand, broadly speaking the term *bifurcation* describes a qualitative change in the orbit structure of a dynamical system (a flow or a map), as one or more of the parameters on which it depends is changed slightly. For example, we have a bifurcation when, owing to a variation in those parameters, the number or the stability properties of fixed points (or limit cycles) of the system change.

Thus, bifurcation theory could be viewed as a study of structurally *unstable* dynamical systems. One might be tempted to comment that bifurcations are unlikely events and therefore practically unimportant. In a sense, this may be true for *individual* given systems, for which bifurcations can be perturbed away by arbitrarily small changes in their structure. It is not so, however, for *families* of systems depending on certain parameters. To illustrate this point we need again the concept of codimension.

From the definition given in the preceding section, it follows that the codimension of a submanifold N of a manifold M is equal to the minimum dimension of a submanifold $P \subset M$ that intersects N in a transversal way. Thus, codimension is a measure of the avoidability of N as one moves about the ambient space.

This geometrical connotation of codimension can be extended to the case of infinite-dimensional manifolds in infinite-dimensional spaces. In the present case, we are concerned with the infinite-dimensional space of dynamical systems. A *bifurcation point* in this space corresponds to a dynamical system which is structurally unstable, due to the occurrence of a singularity in a certain function. (This may correspond, for example, to a vanishing eigenvalue of the Jacobian matrix at equilibrium.) The set of all bifurcation points, F, is a codimension one submanifold in the space of dynamical systems.

On the other hand, a one-parameter family of dynamical systems corresponds to a one-dimensional submanifold (broadly speaking, a curve), γ, in the space of dynamical systems. Generically, γ will intersect F transversally, that is to say, moving along γ by changing the relevant parameter, typically we shall pass through F.

There may of course exist different degrees of non-genericity and the set corresponding to non-generic (structurally unstable) systems may in turn be subdivided into (one-parameter) generic and non-generic subsets (an example of the latter being the intersection of two surfaces corresponding to the simultaneous occurrence of two singularities). Higher-degree non-generic subsets are codimension $2, \ldots, k$ submanifolds in the space

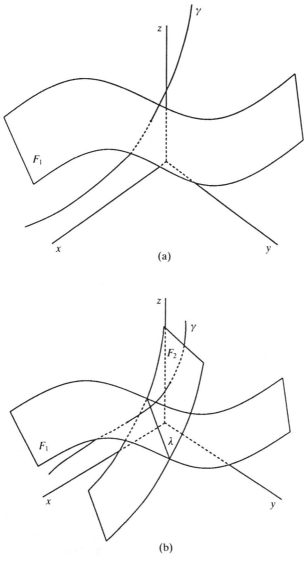

Fig. 2.15. (a) A codimension one bifurcation; (b) a codimension two bifurcation.

of dynamical systems, and they will not be avoidable within $2,\ldots,k$-parameter families of dynamical systems.

Thus, in general, by *codimension k bifurcation*, we indicate the minimum dimension of the parameter space which contains the bifurcation in a

persistent way. A simple example will help clarify the issue. Consider a
continuous-time dynamical system, with three parameters:

$$\dot{x} = f(x; \mu), \tag{2.19}$$

$x \in \mathbb{R}^n$, $\mu = (\mu_1, \mu_2, \mu_3) \in \mathbb{R}^3$. In general, the locus of points in the space
(μ_1, μ_2, μ_3) in which a certain condition is satisfied (e.g., the condition
that one eigenvalue, or the real part of a pair of complex conjugate
eigenvalues of the Jacobian matrix, vanishes), will be a two-dimensional
surface $F_1(\mu_1, \mu_2, \mu_3) = 0$. If we now consider a curve γ in the parameter
space, this will correspond to a one-dimensional family of systems. If γ is
transversal to F_1, which is the generic case, any slightly perturbed curve
γ' will be transversal too. Consequently, although each individual system
(2.19) can be perturbed away from $F_1 = 0$ by arbitrarily small changes,
the γ-family of such systems cannot. (See Fig. 2.15a.)

Suppose now a second, additional condition is required (e.g., that
another eigenvalue, or the real part of a second pair of complex conjugate
eigenvalues of the Jacobian matrix, vanishes). This condition will define a
second, two-dimensional surface $F_2(\mu_1, \mu_2, \mu_3) = 0$, and the simultaneous
verification of F_1 and F_2 will generically define a curve λ in the (μ_1, μ_2, μ_3)
parameter space. Now, generically, an arbitrary curve γ will not intersect
λ in $\mathbb{R}^3$ but a two-dimensional surface will. (See Fig. 2.15b.) That is to
say, we need to move over a parameter space of two dimensions in order
to verify the conditions of a codimension two bifurcation.[34]

The study of bifurcations in dynamical systems of high dimensions may
be very complicated, even when the phenomenon under investigation is
essentially simple. Fortunately, there are techniques that allow one to
systematically reduce the dimension of the state space we need to consider
when analysing bifurcations of a given type. These techniques are based
on the idea of 'centre manifold', i.e., an invariant manifold tangent to
the centre eigenspace. As we know, this is the space spanned by those
eigenvectors of the Jacobian matrix at an equilibrium point which are
associated with eigenvalues with vanishing real parts (or with modulus
equal to one in the case of maps). It can be shown that the local
behaviour of a bifurcating system can be investigated by studying the
dynamics of a 'reduced' nonlinear vector field (or a reduced map) on the
centre manifold.[35]

---

[34] Notice that the dimension of the state space of the system is not relevant here, only the
parameter space and the nature of the bifurcation.

[35] For a detailed analysis of centre manifolds, see Guckenheimer and Holmes (1983,
pp. 123–38).

Thus local bifurcations of codimension one can be studied by means of 'canonical' nonlinear systems involving only one state variable, except for the Hopf bifurcation which requires a state space of dimension two.

Bifurcations can be divided into *continuous* (or *subtle*) and *discontinuous* (or *catastrophic*) ones. Let us call *attractrix*[36] the ensemble of attracting sets of the system

$$\dot{x} = f(x; \mu), \qquad x \in \mathbb{R}^n, \mu \in \mathbb{R}. \tag{2.20}$$

Suppose that (2.20) undergoes a bifurcation at a critical value of the parameter $\mu = \mu_c$.

We call that bifurcation *continuous* if a path in the $(x; \mu)$ control phase-space, parameterized by μ, can be followed continuously across the bifurcation point μ_c, without leaving the attractrix. We call the bifurcation *discontinuous* if the path defined above is not continuous (an analogous definition can be used for maps).[37]

Bifurcations can also be divided into local, global and local/global. We have a *local* bifurcation when the qualitative changes in the orbit structure can be analysed in a neighbourhood of a single fixed point in the phase space (bifurcations of limit cycles can be similarly treated by means of Poincaré maps). *Global* bifurcations proper are characterized instead by changes in the orbit structure of the system which are not accompanied by changes in the properties of its fixed points. *Local/global* bifurcations occur when a local bifurcation (of a catastrophic kind) has global repercussions which qualitatively change the orbit structure far from the bifurcating point.

In what follows we shall almost exclusively consider codimension one bifurcations. Fig. 2.16 (taken from Thompson and Stewart, 1986, pp. 116–17, figure 7.3) provides an overview of the main types of such local bifurcations of flows and maps which are usually encountered in applications. The subject of global bifurcations will be taken up again in Chapter 9, when we discuss transition to chaos.

2.7.1 Hopf bifurcation for flows

The study of Hopf bifurcation plays a particularly important role in the investigation of systems described by ordinary differential equations.

---

[36] This term is Thompson and Stewart's (1986, p. 257).

[37] An interesting situation, akin to bifurcation and sometimes encountered in the literature, is the so-called *canard*, which occurs when the orbit structure of a system changes very rapidly, but continuously, when one or more parameters move across an extremely narrow range of values.

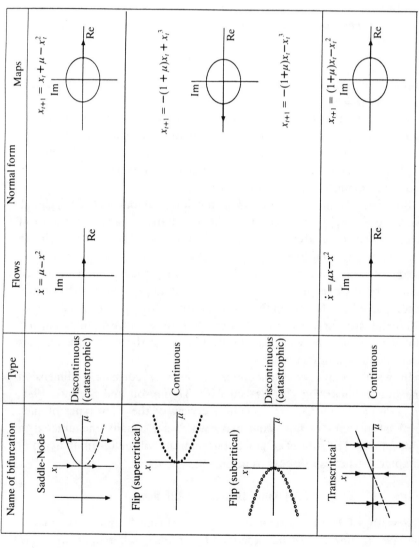

Fig. 2.16. Table of the main codimension one bifurcations.

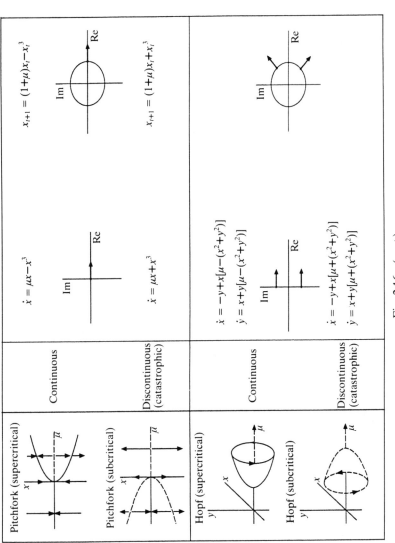

Fig. 2.16. (cont.)

The Hopf bifurcation is one of the two *typical* ways in which systems in equilibrium can lose their stability when one control parameter is changed (the other being the fold bifurcation). The loss of stability of a fixed point and the corresponding birth of a limit cycle is often (though not always) the first step in a 'route to chaos' of a continuous-time system, parameterized by one single control variable.

A celebrated theorem by Hopf, which generalizes certain earlier results of Poincaré and Andronov, provides a rigorous characterization of the dynamics of the system near the bifurcation point and states the conditions for stability of the periodic orbits. There exist several, more or less equivalent versions of this theorem. We shall present here a formulation taken from Invernizzi and Medio (1991).

Theorem 2.4.   *Consider an autonomous ordinary differential equation*

$$\dot{x} = f(x; r), \qquad x \in U, r \in \mathbb{R}, \tag{2.21}$$

where U is an open subset of $\mathbb{R}^n$ and r is a real scalar parameter varying in some open interval $I \subseteq \mathbb{R}$. Suppose that f is a C^5-function $U \times I \to \mathbb{R}^n$, and that for each r in I there is an isolated equilibrium point $\bar{x} = \bar{x}(r)$. Assume that the Jacobian matrix of f with respect to x, evaluated at $(\bar{x}(r); r)$, has a pair of simple complex conjugate eigenvalues $\lambda(r)$ and $\bar{\lambda}(r)$ such that, at the critical value r_H of the parameter, we have

$$Re\ \lambda(r_H) = 0, \qquad Im\ \lambda(r_H) \neq 0, \qquad \frac{d}{dr} Re\ \lambda(r_H) \neq 0,$$

while $Re\ \rho(r_H) < 0$ for any other eigenvalue ρ. Then (2.21) has a family of periodic solutions. More precisely:

(i) there exists a number $\epsilon_0 > 0$ and a C^3-function $r :]0, \epsilon_0[\to \mathbb{R}$, with a finite expansion

$$r(\epsilon) = r_H + \mu_2 \epsilon^2 + O(\epsilon^4),$$

such that there is a periodic solution $x_\epsilon = x_\epsilon(t)$ of

$$\dot{x} = f(x; r(\epsilon)). \tag{2.22}$$

The coefficient μ_2 can be computed by means of certain well-known algorithms (see, for example, Hassard et al., 1981).
If $\mu_2 \neq 0$, then

(ii) there exists a number ϵ_1, $0 < \epsilon_1 \leq \epsilon_0$, such that $r = r(\epsilon)$ is bijective from $]0, \epsilon_1[$ either onto $J =]r_H, r(\epsilon_1)[$ when $\mu_2 > 0$, or onto $J =]r(\epsilon_1), r_H[$ when $\mu_2 < 0$; therefore, the family $x_\epsilon = x_\epsilon(t)$ (for $0 < \epsilon < \epsilon_1$) of periodic solutions of (2.22) can be re-parameterized by the original parameter r as $x = x(t; r)$.

Moreover,

(iii) the stability of $x = x(t;r)$ depends just on a single Floquet exponent $\beta = \beta(r)$, which is a C^2-function of $|r - r_H|^{1/2}$, and it has a finite expansion

$$\beta = \beta_2 \epsilon^2 + O(\epsilon^4)$$

with $\beta_2 = -2\mu_2(d/dr)\text{Re } \lambda(r_H)$, and $\epsilon^2 = (r - r_H)/\mu_2 + O(r - r_H)^2$; thus, for r near r_H, $x(t;r)$ is orbitally asymptotically stable if $\beta_2 < 0$, but it is unstable when $\beta_2 > 0$. Observe that $\beta_2 \neq 0$.

Finally,

(iv) the period $p = p(r)$ of $x(t;r)$ is a C^2-function of $|r - r_H|^{1/2}$, with a finite expansion

$$p = \frac{2\pi}{|Im \ \lambda(r_H)|}(1 + \tau_2\epsilon^2 + O(\epsilon^4)),$$

where ϵ^2 is as in (iii) and τ_2 is also a computable coefficient.

Remark 2.9.   The Hopf bifurcation theorem is *local* in character and only makes predictions for regions of the parameter and phase space of unspecified size.

Remark 2.10.   The sign of the coefficient β_2 (the 'curvature coefficient'), on which stability of the limit cycle depends, is rather hard to check in any system of dimension larger than two. It requires heavy calculations involving second and third partial derivatives. This is particularly unfortunate in economic applications where often only certain general properties of the relevant functions are known (e.g., typically, one knows only that the relevant function is concave).

Remark 2.11.   For $\beta_2 < 0$, the Hopf bifurcation is called *supercritical* and is of a continuous type. For $\beta_2 > 0$, the bifurcation is called *subcritical* and is of a catastrophic type. In the latter case the cycles are not observable and they can be interpreted as 'stability thresholds' of the equilibrium point.

2.7.2 Hopf bifurcation for maps (or Neimark bifurcation)

Equilibrium points of two-dimensional maps can lose their stability when a pair of complex conjugate eigenvalues cross the unit circle, i.e., when their modulus becomes greater than one. When a map is derived from a flow by means of a Poincaré surface of section, the bifurcation of the map corresponds to the loss of stability of a *limit cycle* of the flow.

This phenomenon is often referred to as *secondary Hopf bifurcation* and can be generalized to the bifurcation of an *n*-torus into a $(n + 1)$-torus. Sometimes, however, the limit cycle can bifurcate into another periodic orbit of period kT, T being the period of the original cycle and $k > 2$. The term *Neimark* (or *Neimark–Sacker*) *bifurcation* is also used.

The analysis is more complicated than in the case of flows, but some rigorous, interesting results exist for a certain class of problems. In particular for two-dimensional maps, we have the following theorem (cf. Withley, 1983, pp. 194–9).

Theorem 2.5.   *Let G_μ be a one-parameter family of maps of $\mathbb{R}^2$, which has a smooth family of fixed points $\bar{x}(\mu)$ at which the eigenvalues are complex conjugates $\lambda(\mu)$, $\bar{\lambda}(\mu)$. Suppose:*

(i) there exists a critical value $\mu = \mu_c$ such that $|\lambda(\mu_c)| = 1$;

(ii)

$$\frac{d|\lambda(\mu)|}{d\mu}\bigg|\mu = \mu_c > 0;$$

(the inequality sign is conventional and could be reversed);

(iii) $\lambda^i(\mu_c) \neq 1$ for $j = 1, 2, 3, 4$.

Then there is a smooth μ-dependent change of co-ordinates bringing G_μ into the form:

$$G_\mu(X) = F_\mu(X) + O(|X|^5) \qquad for \ X \in \mathbb{R}^2,$$

and there are smooth functions a, b and θ so that in polar co-ordinates we have:

$$F_\mu(r, \theta) = (|\lambda(\mu)|r + a(\mu)r^3 + \theta(\mu) + b(\mu)r^2).$$

Moreover, for all sufficiently small positive (negative) μ, G_μ has an attracting (repelling) invariant circle if $a(\mu_c) < 0$ (resp., $a(\mu_c) > 0$).

The coefficient a corresponds to the 'curvature coefficient' in the case of flows and, in general, it is equally hard to evaluate (although in the present, two-dimensional case, the calculations are not too difficult).

The theorem above states the conditions for an attracting circle, invariant under the map G, but does not address the problem of describing the dynamics of the map restricted to the circle. There are two basic possibilities according to whether the map G on the circle has periodic orbits or not. To distinguish between the two, we need the concept of *rotation* (or *winding*) *number*. Let us denote the circle by S^1 and consider the homeomorphism $G: S^1 \to S^1$. Making use of the projection

$\exp: \mathbb{R} \to S^1, \theta \mapsto e^{i2\pi\theta}$, we can define an auxiliary map $\bar{G}: \mathbb{R} \to \mathbb{R}$.[38] Then we call the rotation number of the map G the limit:

$$\rho(G) = \lim_{n\to\infty} \frac{\bar{G}^n(\theta) - \theta}{n}.$$

Two cases are possible, namely: (i) $\rho(G)$ is rational (i.e., $\rho(G) = p/q$, where p and q are two integers); then G restricted to the invariant circle has a periodic orbit of period q. (ii) $\rho(G)$ is irrational; then G has no periodic points and the orbit of each point of the circle is dense in it. If we regard G as a Poincaré map of a flow, the case (ii) corresponds to a quasiperiodic orbit of the flow.

How likely are the two cases above? The answer to this question requires a rather subtle argument which we shall omit here. Broadly speaking, we have a situation similar to that occurring in the 'chaotic region' of one-dimensional maps (see below, Chapter 9). Picking a value of the parameter μ at random and examining the dynamics of G_μ on the invariant circle, *typically* we should see periodic orbits.[39] However, the probability of choosing a value of μ for which $\rho(G)$ is irrational and the orbit on the circle is aperiodic is positive.

A more complete picture of the dynamics of the map *after* a Hopf bifurcation (as well as the case of *strong resonance* in which $\lambda(\mu_c)$ is a qth root of unity for $q = 1, 2, 3$ or 4), requires the study of a two-parameter family of maps. Since this question involves the discussion of global bifurcations and transition to chaos, we shall postpone it to Chapter 9.

Remark 2.12. Only saddle-node, flip and Hopf are generic, codimension one bifurcations. All the other types can be perturbed away by introducing an arbitrarily small change in the relevant equations.

Remark 2.13. Notice that bifurcations of fixed points of maps can be interpreted as bifurcations of *cycles* of flows associated with them. In the case of the flip bifurcation the stationary solution is oscillatory, i.e., for any given value of the parameter μ, the solution alternates between two values of the representative variable. This corresponds to a limit cycle of period T of a flow bifurcating into another cycle of period $2T$.

---

[38] In other words, points on the invariant circle of radius r are identified by the angle formed by the abscissa and the segment joining the origin and the point. The maps G and $\bar{G}$ are topologically equivalent in a sense that has been made precise in section 2.6.

[39] In other words, the set of values of the parameter μ yielding rational rotation numbers is open and dense.

2.8 ERGODIC THEORY OF CHAOS

We have so far considered dynamical systems mainly from a geometrical, or topological point of view. However, when the dynamics under investigation are very complex or chaotic, a precise geometrical description of the time evolution of the system may be a very difficult, or impossible task. Nevertheless, statistical information concerning the probability of different types of behaviour might be obtainable. For this purpose, it is sometimes convenient to move from a geometrical approach, centred on the notion of attractor, to a metric or ergodic approach, based on the concept of invariant measure.

Ergodic theory is a difficult and deep subject and here we cannot but briefly mention a few basic ideas relevant to our discussion.[40]

2.8.1 Probability space. Invariant measure

The basic concept here is that of *probability space* which we shall denote by the triplet $(\Omega, \mathscr{F}, \mu)$. Ω is the *sample space*, i.e., the space of points ω designating the elementary outcomes of an experiment. $\mathscr{F}$ is the σ-field (or σ-algebra)[41] of events.

An *event* is a set $A \subset \Omega$ which is of interest. The σ-field $\mathscr{F}$ is the ensemble of all events, i.e., $A \in \mathscr{F}$. Necessarily $\mathscr{F} \subseteq \mathscr{P}(\Omega)$, the power set of Ω, i.e., the set of all subsets of Ω.

In the triplet above, μ designates a *probability measure* on $\mathscr{F}$. The meaning of the latter concept must be made more precise. We first define a *set function* as a real-valued function μ defined on some class of subsets of Ω. A *measure* is a non-negative (possibly infinite) and countably additive set function. More precisely, a set function μ defined on a σ-field $\mathscr{F}$ of subsets of Ω is a *measure* if the following conditions hold:

(i) $\mu(A) \in [0, \infty]$ for $A \in \mathscr{F}$;

(ii) $\mu(0) = 0$;

(iii) if $A_1, A_2, \ldots$ is a disjoint sequence of $\mathscr{F}$-sets, then $\mu(\bigcup_{k=1}^{\infty} A_k) = \sum_{k=1}^{\infty} \mu(A_k)$. (This condition is called *countable additivity*.)

The measure μ is finite or infinite if, for $A \in \mathscr{F}$, $\mu(A) \leq \infty$. If μ is finite and $\mu(\Omega) = 1$ (which can always be obtained by suitable re-scaling), then μ is a *probability measure*. If for an $\mathscr{F}$-set A, the measure of its

---

[40]   The interested reader can consult, for example, Billingsley (1965) and (1979), and Cohen (1980). An excellent introduction to ergodic theory of chaos is provided by Eckmann and Ruelle (1985); for a very useful, short introduction, see also Ruelle (1989, part II).

[41]   A σ-field, or σ-algebra is a class of sets closed under the formation of complements and countable unions.

complement $\mu(A^c) = 0$, then we shall say that A is a *support* of μ and μ is concentrated on A. For a finite measure, A is a support if and only if $\mu(A) = \mu(\Omega)$.

A measure μ is said to be *discrete* if it is concentrated on finitely, or countably many points. On the contrary, if μ is zero on points $x \in A$ (A being the support of μ), we say that μ is *continuous, or non-atomic*. In this case, A is an uncountable set.

Let us now consider a trasformation $T : \Omega \to \Omega$ that is *measurable* in the sense that $[A \in \mathscr{F}] \Rightarrow [T^{-1}A = \{\omega : T\omega \in A\} \in \mathscr{F}]$. If $\mu(T^{-1}A) = \mu(A), \forall A \in \mathscr{F}$, then T is said to be *measure-preserving* (and μ is said to be *invariant* under T).[42]

When these concepts are applied to the investigation of dynamical systems the sample space Ω will typically correspond to the phase space, the elements ω of Ω to states, or positions of the system. The subsets A (the events) will correspond to certain interesting configurations of orbits of the system in the phase space, such as fixed points, limit cycles or strange attractors. Finally, the transformation T will correspond to the (flow) map governing the motion of the system.[43]

Now let $(\Omega, \mathscr{F}, \mu)$ be a probability space, let f be a μ-integrable real function, and let μ be invariant with respect to a map T. Then the following limit exists with probability 1 (i.e., μ-almost everywhere):

$$\lim_{n \to \infty} \frac{1}{n} \sum_{k=0}^{n-1} f(T^k \omega) = \hat{f}(\omega) \qquad (2.23),$$

where $\hat{f}(\omega)$ is an invariant function such that

$$\int_\Omega f(\omega) d\mu(\omega) = \int_\Omega \hat{f}(\omega) d\mu(\omega). \qquad (2.24)$$

The expressions in equation (2.24) can be identified as conditional expected values of $f(\omega)$ and $\hat{f}(\omega)$, respectively, with respect to the σ-field of invariant sets (which is a σ-subfield of $\mathscr{F}$). The equation says that these expected values are equal. Heuristically, we can say that this statement – which is known as the Birkhoff–Khinchin theorem – belongs to the class of 'laws of large numbers'. It implies that the presence of an invariant measure makes it possible to take averages over time (or over iterations of a map).

---

[42] If T is one-to-one, $T\Omega = \Omega$ and $[A \in \mathscr{F}] \Rightarrow [TA = \{T\omega : \omega \in A\} \in \mathscr{F}]$, then T is said to be *invertible*.

[43] Since, in the present context, there is no essential difference between discrete- and continuous-time dynamical systems, in what follows we shall discuss the issue mostly in terms of maps, i.e., discrete-time systems.

2.8.2 Ergodicity

The second important property to consider here is *ergodicity*. Heuristic-ally, we shall say that a transformation T is *ergodic* (or *indecomposable*, or *metrically transitive*) if it has the property that, for almost every ω, the orbit $\{\omega, T\omega, T^2\omega, \ldots\}$ of ω is a sort of replica of Ω itself.

Formally, we shall say that T is ergodic if each invariant set A, i.e., a set such that $T^{-1}A = A$,[44] is trivial in the sense that it has measure either zero or one. The transformation T is ergodic if in the Birkhoff–Khinchin theorem, for any integrable, real-valued function f, the limit value $\hat{f}$ is constant and we have μ-a.e.

$$\lim_{n\to\infty} \frac{1}{n} \sum_{k=0}^{n-1} f(T^k\omega)d\mu(\omega) = \hat{f}(\omega) = \int_\Omega f(\omega)d\mu(\omega).$$

In this case, the average value of $f(\cdot)$, evaluated along the orbit $T^k\omega$, converges μ-almost everywhere to the mathematical (conditional) expec-tation, or mean of $f(\cdot)$, evaluated on the space Ω. This situation is often described by saying that, for ergodic systems, the time average is equal to the space (or phase) average.

Let us now set $f = I_A$, where

$$I_A(\omega) = \begin{cases} 1, & \text{for } \omega \in A, \\ 0, & \text{for } \omega \notin A \end{cases}$$

is called the *indicator* (or *characteristic function*) of A. Thus,

$$\sum_{k=0}^{n-1} I_A(T^k\omega)$$

is the number of elements of the sequence $\{\omega, T\omega, T^2\omega, \cdots\}$ that lie in A. If the sequence is identified as an orbit of a dynamical system, the sum above indicates the number of times the orbit visits the set A. Although we expect in general this sum to be infinite, the average time spent in A by the orbit can be finite. Indeed, if T is ergodic and $\mu(A) > 0$, it can be shown that

$$\lim_{n\to\infty} \frac{1}{n} \sum_{k=0}^{n-1} I_A(T^k\omega) = \int_\Omega I_A(\omega)d\mu(\omega) = \int_A d\mu(\omega) = \mu(A),$$

μ-almost everywhere.

Hence, the orbits of an ergodic system replicate the sample space Ω in the sense that, for every measurable set $A \subset \Omega$, an orbit starting from μ-

---

[44] Or, if T is invertible, $TA = A$.

almost every point ω (i.e., the typical orbit) 'visits' A with asymptotic frequency $\mu(A)$.

2.8.3 Absolute continuity and observability

The result just mentioned may not be sufficient to infer that a set A which supports a measure μ such that $\mu(A) > 0$ is actually relevant from a physical (economic) point of view. Indeed the set defined by the expression 'μ-almost every point $\omega \in \Omega$' may still be too small in relation to the phase space of the dynamical system in question. A physically more natural and relevant notion of sampling is given by the *Lebesgue measure*, henceforth denoted by m. If we consider the unit interval together with the so-called σ-field B of Borel sets,[45] this is the probability measure defined by the requirement that

$$m(a, b] = |b - a|,$$

for subintervals of $(0, 1]$. The Lebesgue measure is the unique probability measure on B that assigns to each interval its length as its measure. There also exists a k-dimensional space analogous to the Lebesgue measure which assigns to bounded rectangles in $\mathbb{R}^k$ their k-volumes.

Consider now the following case: a dynamical system is characterized by an invariant set A which supports an ergodic, invariant, continuous measure μ such that $\mu(A) > 0$, but, at the same time, the set of points attracted by A has Lebesgue measure zero, i.e., it is negligibly small in relation to the sample space Ω (in our case the phase space). Then, the motion of the system on the invariant set would not be observable, e.g., it would not be possible to show it on the screen of a computer. In this case, the orbit that visits the set A with a positive asymptotic frequency $\mu(A)$ is 'typical' with respect to μ, but 'exceptional' with respect to m. If so, the set A cannot be considered an attractor of the system in a physically relevant sense.[46]

When the situation just depicted occurs, we would say that the measures μ and m on $(\Omega, \mathscr{F})$ are *mutually singular*, i.e., they have disjoint supports. That is to say, there exist sets S_μ and S_m such that

$$\mu(\Omega - S_\mu) = 0, \quad m(\Omega - S_m) = 0, \quad S_\mu \cap S_m = 0.$$

In order to guarantee that such a somewhat paradoxical situation does

---

[45] If $\mathscr{F}$ is the class of sub-intervals of $\Omega = (0, 1], B \equiv \sigma(\mathscr{F})$ is the intersection of all σ-fields containing $\mathscr{F}$. The elements of B are called the *Borel sets* of the unit interval.

[46] Chaos which is not observable in the sense just discussed is sometimes referred to as 'thin' (or 'formal', or 'topological') chaos, whereas observable chaos is nicknamed 'thick' chaos.

not occur, we must verify that the relevant measure μ is *absolutely continuous with respect to* m (for short, *absolutely continuous*), or equivalently, that μ is *dominated by* m (sometimes denoted by $\mu \ll m$).

A finite measure μ is said to be absolutely continuous with respect to the Lebesgue measure m if there exists a non-negative integrable function $f(\cdot)$ such that $\mu(A) = \int_A f\,dm$ for all measurable sets A. The function f is called the density of μ.

Absolute continuity of μ implies that, for all measurable sets A, $[m(A) = 0] \Rightarrow [\mu(A) = 0]$ (and therefore $[\mu(A) > 0] \Rightarrow [m(A) > 0]$). Consequently we can say that, for example, a chaotic invariant set A with an absolutely continuous measure $\mu(A) > 0$ represents the asymptotic behaviour of solutions for a non-trivial set (i.e., a set with positive Lebesgue measure) of initial conditions and is therefore observable.

In applications, and especially in economic applications, the concept of absolute continuity of measures has mostly been referred to one-dimensional maps, i.e., to dynamical systems whose motion can take place in only one direction. However, when we consider higher dimensional maps or flows characterized by chaotic dynamics, we do not expect invariant measures to be smooth in all directions. On the contrary, given the complicated structures of the orbits of these dynamical systems, the measures may have rough densities in the directions transversal to the foliations of the unstable manifolds.

In this case, the observability of a chaotic invariant set may be established by proving the existence, for the dynamical system in question, of an invariant ergodic measure μ which is absolutely continuous *along the unstable directions*. Measures having this property are known as SRB (Sinai, Ruelle, Bowen) measures.[47]

The existence of an SRB measure μ is of particular interest in the present context since, at least in a number of cases, it can be shown that the physical (or computer-generated) time averages tend to μ not just for μ-almost all points in Ω (which might in the general case be a

---

[47] More precisely, SRB measures can be defined as follows (cf. Ruelle, 1985, pp. 73–4). Consider a μ-measurable set of the form $S = \bigcup_{\alpha \in A} S_\alpha$, where the S_α are disjoint pieces of unstable manifolds W^u. Then, if the decomposition is μ-measurable, we can define 'μ restricted to S' as

$$\int \mu_\alpha \rho(d\alpha),$$

where ρ is a measure on A and μ_α is a probability measure on S_α called a 'conditional probability measure'. Then we say that an ergodic measure μ is SRB if its conditional probabilities μ_α are absolutely continuous with respect to the Lebesgue measure for some choice of S with $\mu(S) > 0$.

negligibly small set), but for a set of positive Lebesgue measure. Thus, invariant chaotic sets characterized by a SRB measure can be said to be observable in a physically relevant sense.

2.8.4 Some examples of invariant measures

At this point, it is perhaps expedient to consider some examples of measures associated with the most common types of invariant sets discussed in the previous sections.

(a) Attracting fixed point of a flow

If $\bar{\omega} \in \Omega$ is a point such that $T^k\bar{\omega} = \bar{\omega}$, for all k, then

$$\mu = \delta_{\bar{\omega}}$$

(where $\delta_{\bar{\omega}}$ is the Dirac delta function centred at $\bar{\omega}$), is an invariant, ergodic measure. The measure is, of course, discrete.

(b) Attracting k-periodic point of a map

Here the steady state of the time evolution of the system is a sequence $\{\omega, T\omega, \ldots, T^k\omega = \omega\}$. In this case, the (discrete) measure that assigns $1/k$ of the mass to each point of the sequence is invariant and ergodic.

(c) Attracting limit cycle of a flow

Here we have $T^k\omega = \omega$ for some positive k and the cycle is a closed orbit defined by $\gamma = \{T^t\omega | t \in [0, k)\}$. In this case, the (continuous) measure

$$\mu = \delta_\gamma = \frac{1}{k}\int_0^k \delta_{T^t\omega}dt$$

is invariant and ergodic.

(d) Quasiperiodic attractor (of a flow)

In the simple two-dimensional case, the steady state time evolution of the system on a T^2 torus can be described by the vector field

$$\begin{aligned}\dot{\theta}_1 &= \omega_1, \\ \dot{\theta}_2 &= \omega_2,\end{aligned} \tag{2.25}$$

where θ_1, θ_2 are suitable angular variables. As we have already mentioned, if the ratio between the two frequencies, ω_1/ω_2, is irrational, the orbits will be quasiperiodic and dense in T^2. However, the orbits will not be chaotic in the sense of SDIC (i.e., orbits starting close will remain close).

In this case, it can be shown that the (Haar) measure

$$\frac{d\theta_1 d\theta_2}{(2\pi)^2}$$

is invariant under the flow and is ergodic.

(e) Quasiperiodic attractor (of a map)

A simple example of this case is given by the so-called circle map

$$G: S^1 \to S^1,$$
$$\theta \mapsto G(\theta) = \theta + 2\pi\alpha \ (\text{mod } 2\pi). \tag{2.26}$$

If α is irrational, the orbit of any point of the circle is dense in the circle, i.e., it passes arbitrarily near any other point and is not periodic. There is no SDIC and the motion is not chaotic in this sense. Notice that the map G can be viewed as the Poincaré map of the continuous-time system described above under (d). In this case, α will be equal to the ratio of the two frequencies, ω_1/ω_2.

It can be shown that map (2.26) preserves the so-called circular Lebesgue measure. Roughly speaking, the latter assigns to arcs of the unit circle their lengths as their measure.[48]

(f) Chaotic, strange attractor

In this case, the situation is more complicated and seldom can we actually calculate an ergodic invariant measure.[49]

A noticeable exception is given by the well-known special case of the 'logistic map' (cf. Chapter 9 below)

$$f(x) = 4x(1 - x),$$

which can be shown to preserve the measure

$$\mu = \frac{1}{\pi} \frac{dx}{[x(1 - x)]^{1/2}}$$

(cf. Collet and Eckmann, 1980).

In general, a chaotic attractor may carry uncountably many distinct ergodic measures, and it is not clear which of them one should select. It

---

[48] Let $\mathscr{B}$ consist of the Borel subsets of the interval $[0, 1)$, C be the unit circle in the complex plane, and define $T: [0, 1) \to C$ by $T\omega = e^{2\pi i\omega}$. Then, if m is the ordinary Lebesgue measure on $\mathscr{B}$, the circular Lebesgue measure on $\mathscr{C} = [A : T^{-1}A \in \mathscr{B}]$ is given by mT^{-1}, and is invariant under rotations.

[49] However, there is a theorem (cf. Eckmann and Ruelle, 1985, p. 626) stating that if a compact set A is invariant under a transformation T, then there exists a probability measure μ invariant under T and with support contained in A. One may choose μ to be ergodic.

has been suggested (see, for example, Eckmann and Ruelle, 1985, p. 626; Ruelle, 1989, pp. 27 and 44) that a physically natural choice may be provided by the so-called *asymptotic*, or *Kolmogorov measure*.

From an experimental point of view, the time evolution of actual physical systems seems, in many cases, to produce well-defined time averages, and the same applies to computer-generated time series. If the time evolution is taking place on an invariant set A, those time averages can be taken to define a probability measure μ on A, describing how frequently the various parts of A are visited by the orbit of the system. The probability measure thus defined operationally is called a *physical measure*. Consider now that, owing to the presence of external noise or round-off errors, physical or computer-generated time evolutions can be considered as stochastic processes. The latter normally have only one stationary measure μ_ϵ, where ϵ indicates the level of noise. The Kolmogorov measure is the limit (assuming that this exists) of μ_ϵ when $\epsilon \to 0$. This approach has been shown to be mathematically well-founded for a certain class of dynamical systems (the so-called Axiom-A systems).[50]

An alternative way of selecting an invariant ergodic measure is given by the already mentioned SRB (Sinai, Ruelle, Bowen) measures. As we have seen, the latter can be defined as measures which are absolutely continuous along the stretching, or unstable directions of a dynamical system. SRB measures, when they exist, are operationally given by the physical (or computer-generated) time averages calculated for all initial points x_0 in a subset S of the phase space Ω with Lebesgue measure $m(S) > 0$. It can be shown that, for Axiom-A systems, the Kolmogorov and SRB measures coincide. As we have seen above, SRB measures are also important in relation to the matter of the observability of chaotic attractors.

2.8.5 Entropy and Lyapunov characteristic exponents

As we have seen in the preceding sections, sensitive dependence on initial conditions, although not the only criterion used to characterize chaos, is the essential feature of chaotic dynamics from a physical (economic) point of view. SDIC is at the root of the 'disorderly' behaviour of

---

[50] Let $f : A \to A$ be a diffeomorphism (A being a compact manifold of finite dimension m) and let $B \subset A$ be the non-wandering set for f. Then f is called an *Axiom-A diffeomorphism* if B is hyperbolic (as defined elsewhere in this chapter) and if the periodic points are dense in B (cf. Eckmann and Ruelle, 1985, p. 636).

deterministic dynamical systems and is responsible for their random appearance and unpredictability. The ergodic approach can be used to gain fresh insight on this question and to define the notion of 'disorder' precisely so that chaos may be identified in specific cases. This problem can be discussed in terms of the two related concepts of *entropy* and *Lyapunov characteristic exponents.*

Broadly speaking, if we consider a random variable X taking a finite number of values with probability $p_1, \ldots, p_N$, we define the entropy of X as the quantity[51]

$$H = - \sum_{i=1}^{N} p_i \log(p_i).$$

H measures the uncertainty concerning X, or equivalently the amount of information we get on the average by making an observation.

If the variable is continuous, we can divide the space into intervals I_j $(j = 1, \ldots, p)$, of a given finite size $\leq \epsilon$, and define:

$$H(I) = - \sum_{j=1}^{p} \mu(I_j) \log \mu(I_j),$$

where $I = (I_1, \ldots, I_p)$, and μ is an invariant, ergodic probability measure. When we deal with a dynamical system represented by a map ϕ, the partitioning can be done as follows. Let $J = (J_1, \ldots, J_p)$ be a finite, μ-measurable partition of the support of μ. For each piece J_i, we write $\phi^{-k} J_i$ for the set of points mapped by ϕ^k to J_i, where ϕ^k, of course, means the k-th iterate of ϕ. Then we define the partition:

$$J^{(n)} = \bigcup_{k=1}^{n} \{\phi^{-k}(J)\}.$$

The partition $J^{(n)}$ is deduced from J by time evolution over an interval of length n.

We can then define:

$$H(J) = \sum_{i=1}^{p} \mu(J_i) \log \mu(J_i),$$

and, analogously, $H(J^{(n)})$. Intuitively, $H(J)$ represents the amount of information contained in the partition J, and $H(J^{(n)})$ is the same over a time interval n. If we now let n go to infinity, we can define:

$$h(\mu, J) = \lim_{n \to \infty} \frac{1}{n} H(J^{(n)}),$$

---

[51] The definition of entropy that follows is drawn from Carleson (1989, pp. 2–3) and Eckmann and Ruelle (1985, p. 638).

and

$$h(\mu) = \lim_{d(J) \to 0} h(\mu, J),$$

where $d(J) = \max_i\{\text{size of } J_i\}$.

The quantity $h(\mu, J)$ represents the rate of information creation with respect to the partition J, and $h(\mu)$ – which is called the *entropy* of μ – is the limit of that rate for finer and finer partitions. The quantity $h(\mu)$ is also known as 'Kolmogorov–Sinai entropy'.

Entropy thus defined is often taken as the most relevant disorder parameter in order to identify chaos. In particular, the presence of positive entropy indicates that, for any finite but arbitrarily fine coarse-graining of the space in which the dynamical system in question evolves, the observation of the system continues to generate information for an arbitrarily long interval of time. Consequently, unless the position of the system can be observed with absolute precision, there will forever remain uncertainty about its future course, even when the dynamical rule governing the system is known with precision.[52]

The second, related way of measuring 'disorder' in a dynamical system, is based on the concept of Lyapunov characteristic exponents (LCEs). Since this concept (and a number of related analytical and numerical questions) will be discussed later at some length, we refer the reader to Chapter 6 which is entirely devoted to the subject. We shall limit ourselves here to only a few broad considerations. For this purpose, let us consider the case of a (discrete- or continuous-time[53]) dynamical system living in an n-dimensional phase space. If the initial state of a time evolution of the system is slightly perturbed in one of the n directions, the exponential rate at which the perturbation increases (or decreases) with time is called a (Lyapunov) characteristic exponent. The presence of one or more positive LCEs thus indicates exponential separation of nearby orbits, i.e., sensitive dependence on initial conditions. The latter, as we have seen above, is the essential property of chaotic, or strange attractors.

There is a very nice and interesting relation between the entropy $h(\mu)$,

---

[52] See our discussion of this point in Chapter 1 above. Entropy is also related to the concept of 'information dimension', which will be discussed later in Chapter 7.

[53] As indicated earlier in this chapter, the continuous-time case can be easily reduced to the discrete-time one. If $\phi(t)$ denotes the solution of the system, we can discretize it, for example, by simply taking the corresponding time-one map $G = \phi_1$. None of the present considerations would be altered in any essential way.

as defined above, and the positive characteristic exponents of a dynamical system, denoted here by λ_i^+, namely:

(i) $h(\mu) \leq \sum_i \lambda_i^+$, always;

(ii) equality holds in (i) if and only if μ is absolutely continuous along unstable manifolds.

This means that, in case (i), the sum of the positive exponents fixes an upper bound for entropy and, in the more interesting case (ii), there is equivalence between chaos in the sense of positive entropy, and chaos in the sense of SDIC as signalled by Lyapunov exponents.

Remark 2.14.   Sometimes one reads in the literature that if the invariant measure μ associated with a dynamical system is continuous and ergodic, then the orbits of the system must have a complex, or even chaotic structure. However, this is not quite correct. As we have seen in the examples discussed above, for continuous-time systems, an invariant, ergodic, continuous measure may be associated with orbits no more complex than a limit cycle. For discrete-time systems, the existence of an invariant, ergodic, continuous measure need not indicate the presence of chaotic dynamics, but may be associated with quasiperiodic orbits. Also, the relevance of absolute continuity of measures for the detection of strange attractors is sometimes presented in a misleading manner. As we hope has been made clear by the discussion above, absolute continuity (of an invariant, ergodic measure) does not imply the presence of 'disorder' (positive entropy or SDIC) and therefore cannot in itself signal the presence of chaos; it only concerns the observability of chaos, when the latter exists.

3

A user's guide

In this short chapter we shall provide a few practical suggestions for the application-oriented reader who wants to use this book and the attached program, without following all the mathematical details. We also assume that the reader's main purpose is that of detecting chaotic behaviour in a certain nonlinear system.

The first step is usually to write a model of the system in the form of a set of first-order, ordinary differential (or difference) equations. Not all problems, specifically not all economic problems, can be thus formulated. There are economic problems, for example, optimal growth, which may require the employment of *partial* differential equations, or differential *inclusions*. Economic dynamical questions in general might turn out to be best formulated in terms of *cellular automata*, i.e., by models in which both time and the dependent variables can only take discrete values.[1] In all these cases, difficult logical, mathematical (and computational) questions are involved and we do not deal with any of these less common dynamical models in this book.

After this preliminary step, we usually wish to find out whether the motion of the system (starting from some 'reasonable' initial conditions) is restricted to a subset of the state space, often but not always in the positive orthant. For this purpose, we might like to prove the existence of an attracting set or trapping region, i.e., a region of the state space which attracts all the orbits starting in a certain neighbourhood, and such that the system never leaves it. This may be done, for example, by defining an appropriate Lyapunov function and showing that its derivative with respect to time is negative everywhere in a certain region of the space. Sometimes the existence of an attracting bounded region cannot be proved analytically and we have to rely on numerical evidence of its presence.

---

[1] On cellular automata and related subjects, see Jackson (1990, vol. II, pp. 436ff.).

The third step is to prove the existence of one or more attractors within the attracting region. The simplest result is that of finding equilibrium (or fixed) points. This involves locating the solution(s) of a set of nonlinear algebraic equations obtained from the original system by putting all the derivatives w.r.t. time equal to zero (or, in the case of maps, by putting $x_t = \bar{x}, \forall t$). No general straightforward methods exist to find all such solutions, but in many applications the problem can be solved easily. Local or global stability of fixed points may then be ascertained by one or another of the methods discussed in Chapter 2 above.

Periodic solutions of discrete-time models (maps) correspond to fixed points of iterates of the maps, and their investigation is no more difficult than that of equilibrium points of continuous-time systems. Periodic solutions of systems of differential equations (limit cycles) can be found out by local analysis, making use of the Hopf bifurcation theorem, which also provides us with (not so straightforward) methods to establish their stability.

Other types of attractors are harder to pinpoint. The presence of quasiperiodic solutions can sometimes be established by locating Hopf bifurcations of maps or secondary Hopf bifurcations of flows. The existence of chaotic attractors, however, can be proved analytically only in a small number of rather special cases. Certain theoretical predictive criteria exist which entail delicate global considerations (for example the existence of a homoclinic orbit) and they will be concisely discussed in Chapter 8. The reader's attention is drawn to the fact that, at present, rigorous analytical results concerning strange, or chaotic attractors are restricted to a very small number of specific areas, including one-dimensional non-invertible maps (see Chapter 9), two-dimensional invertible maps of the Hénon type and continuous-time systems of the Lorenz type.[2]

We must conclude, therefore, that at present, complete (though approximate) information on the structure of orbits of continuous dynamical systems of dimension greater than two (or of maps of dimension greater than one) can only be obtained by numerically integrating the equations of the systems, studying their geometry and computing the values of certain crucial quantitative properties. Although theoretical knowledge is not sufficient to provide a complete picture of the dynamics of a

---

[2] For recent results on the Hénon model, see Benedicks and Carleson (1989); on the Lorenz model see Sparrow (1982) and, more recently, Robinson (1988). Notice that in both cases, in spite of the enormous amount of numerical evidence and intellectual investment, the analytical results are still partial.

system, it is nevertheless indispensable to guide numerical computations and to interpret their results. In applications, therefore, and specifically in economics, the most promising research strategy seems to us to be an association of analysis and numerical simulation through which we can establish a series of 'symptoms' whose concurrence is the best available indicator of the presence of those patterns of behaviour we are looking for.

In Chapters 4–9 we have discussed the theoretical characterizations of a number of typical 'symptoms' of chaotic behaviour for the numerical analysis of which our program provides computational facilities.

Suppose we have ascertained that for the system under investigation there exists an attracting set, but the asymptotic motion within it is not (or does not seem to be) a simple one (i.e., a fixed point or a periodic orbit). The first thing is to carry out a visual inspection of (numerically generated) trajectories and a 2D or 3D projection of them on the phase space. It may happen that the trajectories, after a transitory period, appear to settle on an attractor with more or less complicated geometry. Then we would typically like to verify the presence of these 'symptoms':

- irregular, *prima facie* aperiodic, time profile of the trajectories of the system;
- trajectories starting from nearby initial points which become uncorrelated after a (small) finite number of computer iterations;
- rapid decay of the autocorrelation function;
- broad power spectrum of trajectories on the attractor;
- a post-transient first return map which seems to consist neither of one point nor of a finite collection of distinct points;
- a non-integer fractal dimension on the attractor (as calculated, for example, by means of the Grassberger and Procaccia method);
- one or more positive Lyapunov characteristic exponents.

These symptoms can be evaluated in a 'static' manner, i.e., they can be referred to trajectories (apparently) belonging to a given attractor. Their significance, however, can be enhanced by combining this study with a 'stroboscopic' investigation, of the emergence and the evolution of each of those symptoms when certain parameters are changed. This is particularly relevant in ascertaining whether the system is following one of certain known 'routes to chaos', which we discuss in Chapter 9.

As an example, let us suppose that we are studying a family of continuous-time systems depending on one parameter, say

$$\dot{x} = f_r(x), \qquad x \in \mathbb{R}^n, \qquad (3.1)$$

and we want to know how its asymptotic behaviour evolves as we vary the parameter r. A rather common, but by no means unique, scenario is the following:

Stage 1.   For small r, say $r < r_1$, system (3.1) has a unique globally attracting equilibrium point $\bar{x}_1$.

Stage 2.   At $r = r_1$ there is a bifurcation of solutions, so that $\bar{x}_1$ loses its stability and a new equilibrium $\bar{x}_2$ appears, which is stable for, say, $r_1 < r < r_2$.

Stage 3.   At $r = r_2$ a Hopf bifurcation takes place, the equilibrium point $\bar{x}_2$ loses its stability and an initially stable limit cycle of period T is born.

Stage 4.   At $r = r_3 > r_2$ a secondary Hopf bifurcation occurs, and an invariant two-dimensional torus appears on which the solution is quasi-periodic.

Stage 4 bis.   Alternatively, at $r = r_3$, a period-doubling bifurcation occurs, giving rise to a period $2T$ cycle. This step can be repeated an indefinite number of times as we further increase r, leading to a cascade of bifurcations.

Stage 5.   For $r > r_3$ the motion of the system, although still confined to a bounded subset of the phase space, looks irregular and symptoms of chaos, such as a broad band power spectrum or a positive Lyapunov exponent appears.

Stage 6.   For even larger values of r, the system may 'explode' (the computer will show an overflow error), or its dynamics may again become regular (periodic).

Of course, the transition between stage 3 and stage 5 is a most delicate, not completely understood, question and an exhaustive classification of 'routes to chaos' is still lacking. A helpful numerical device to investigate such a transition is the so-called 'bifurcation diagram' by which one broadly means a representation of any characteristic property of the solution of the system as a function of a bifurcation parameter. Bifurcation diagrams are relatively easy to produce for maps, but not so easy, or accurate for flows. Our program provides both, and examples of bifurcation diagrams for continuous-time models can be found below in Chapters 11 and 13.

We would like to conclude this user's guide by mentioning a rather deep and intriguing question concerning the reliability of the computation of chaotic systems. We have already said that trajectories of such systems are characterized by extreme sensitivity to initial conditions. On the other hand, numerical simulations involve a very large number of computations, each of them containing necessarily some errors. We might thus naturally question the validity of numerical investigation of chaotic systems for which, paradoxically, such investigation is particularly important.

Fortunately the situation is not as bad as it looks. Two theorems by Anosov (1967) and Bowen (1978) have proved that, for uniformly hyperbolic invariant sets, numerical (or noisy) orbits will stay close to (be shadowed by) true orbits for all time. Uniformly hyperbolic sets are rarely encountered in applications. However, a recent result by Hammel *et al.* (1987) shows that for general (nonhyperbolic) sets, typically numerical orbits are shadowed by true orbits for long time periods. Thus, according to this result, we expect a calculated orbit to be a good approximation of the true (chaotic) dynamics of the system.[3]

---

[3] The question, however, is still open. For a dissenting view, see McCawley and Palmare (1986).

4

Surfaces of sections and Poincaré maps

4.1 INTRODUCTION

The concepts of 'surface of sections' and 'Poincaré map' refer to a method which reduces the study of continuous-time dynamical systems (flows) to the study of discrete-time systems (maps). The method was introduced by Poincaré at the end of the last century.

The main idea behind it is that, in order to investigate the asymptotic behaviour of a continuous-time dynamical system, it is not really necessary to describe the entire course of its orbits, but it may be sufficient to observe its position at discrete intervals of time which are not necessarily of equal length.

Consider, for example, the case of a differential equation

$$\dot{x} = f(x), \qquad x = x(t) \in \mathbb{R}^n. \tag{4.1}$$

If we take discrete-time samples of the vector-valued variable x, we obtain a sequence of data $\{x_1, x_2, \ldots, x_n\}$, where, by definition, $x_n \equiv x(t_n)$. If, moreover, for any n, the values of x_{n+1} could be determined as a function of x_n, it would be possible to write a map of the form

$$x_{n+1} = g(x_n), \tag{4.2}$$

and study (4.2) instead of (4.1).

There are several justifications for this procedure, namely:

(i)　certain essential dynamical properties of the differential system (4.1) are reflected by map (4.2);[1]

(ii)　the study of (4.2) (an ordinary equation) is simpler than that of (4.1) (a differential equation), both theoretically and numerically;

---

[1] There are, however, a few cases in which differences in dynamical behaviour between systems cannot be recognized by simply looking at their Poincaré maps. See, for example, Hao (1989, p. 280).

(iii)   the derivation of map g entails a reduction of at least one unit of the dimension of the relevant phase space;[2]

(iv)   the graphical representation of iterations of the map is often simpler and more suggestive than the corresponding representation of the differential equation.

The Poincaré surface of section (and the resulting Poincaré map) can be conceived as a special way of choosing sampling times t_n of a (continuous) sequence of variables generated by a flow.

4.2 'CANONICAL' POINCARÉ MAPS

Rigorous analytical methods for constructing Poincaré surfaces of sections and maps are not available generally, i.e., for arbitrary systems of differential equations. However, for some specific classes of such systems, certain canonical procedures have been developed. In particular, this is true of (i) systems characterized by periodic orbits; (ii) systems with periodic or quasi-periodic forcing terms; (iii) systems characterized by homoclinic or heteroclinic orbits[3]. Let us discuss these in turn.

Case (i). Poincaré maps near a periodic orbit

Consider, for example, the nonlinear differential system:

$$\dot{x} = f(x), \qquad x \in \mathbb{R}^n, \tag{4.3}$$

and the related flow $\phi(t, x)$. Let γ be a periodic orbit of period T arising from (4.3), i.e., $\phi(t + T, x_0) = \phi(t, x_0)$, where $x_0 \in \mathbb{R}^n$ is any point through which the periodic solution passes. A *local cross-section* $\Sigma \subset \mathbb{R}^n$ is a hypersurface of dimension $(n - 1)$, chosen in such a manner that the flow is everywhere transverse to it at x_0. Formally, if we define by $n(x_0)$ the unit normal to Σ, we should have everywhere $\langle f(x), n(x_0) \rangle \neq 0$ where $\langle \cdot, \cdot \rangle$ indicates the scalar product operation. If $U \subseteq \Sigma$ is a neighbourhood of x_0, then under certain mild conditions on $f(x)$, we can define a map which associates points in U with their points of first return to Σ in a time close to T. More precisely, we shall have

$$P : U \to \Sigma.$$
$$x \mapsto \phi(\tau(x), x), \tag{4.4}$$

where τ is the time of first return of the point x to Σ. In general $\tau = \tau(x)$

---

[2] A dimensional reduction of a problem by more than one unit may be possible, for example, when the phase-space is made up of families of tori. See Wiggins (1990, p. 82).

[3] Cf. Wiggins (1988, pp. 67–8), Guckenheimer and Holmes (1983, pp. 22–7).

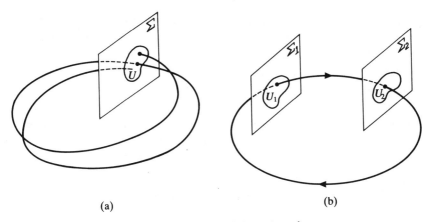

(a)   (b)

Fig. 4.1. The geometry of the Poincaré map.

and it is not necessarily equal to the period $T = \tau(x_0)$ of the periodic orbit γ. However, $\tau \to T$ as $x \to x_0$.[4] Map (4.4) is called the Poincaré map and it is illustrated in Fig. 4.1.

It can be readily seen that x_0 is an equilibrium, or fixed point, of map (4.4) and corresponds to a periodic orbit of period T of the flow. Moreover, a periodic point of period $k > 1$[5] for the map corresponds to a subharmonic of period kT for the flow. Moreover, the stability properties of x_0 for the map P reflect the corresponding properties of the periodic orbit γ for the flow $\phi(t, x)$. This can be explained by first writing a solution of (4.3) lying on the closed orbit γ emanating from $x_0 = p$, as $\bar{x}(t) = \bar{x}(t + T)$, T being again the period of the orbit. Next, we linearize (4.3) about γ and write

$$\dot{\xi} = Df(\bar{x}(t))\xi, \tag{4.5}$$

where $\xi \equiv x - \bar{x}$ and $Df(\bar{x}(t))$ is a time-dependent $n \times n$, T-periodic matrix.

The behaviour of solutions of (4.3) in the neighbourhood of γ depends on the eigenvalues of a constant $n \times n$ matrix e^{TR} which can be determined (uniquely but for a similarity transformation) by the solutions of (4.5).

---

[4] Notice that the properties of the Poincaré map do not depend on the choice of a particular surface of section Σ. It can be proved that any two different maps $P_1 : U_1 \to \Sigma_1$ and $P_2 : U_2 \to \Sigma_2$, constructed according to the requirements specified above, are locally conjugate, and therefore their dynamics are locally the same.

[5] I.e., a point $x \in U$, such that $P^k(x) = x$, but $P^j(x) \neq x$ for $1 \leq j \leq k - 1$, where of course P^k means P iterated k times.

The n eigenvalues of e^{TR} are also called *characteristic Floquet multipliers* of the close orbit γ. It can be shown that (i) the multiplier associated with perturbations along the orbit γ is always equal to one and (ii) the remaining $(n-1)$ multipliers are the eigenvalues of the Poincaré map, linearized around the fixed point, i.e., $DP(x_0)$. Consequently, the stability of the orbit γ can be ascertained by calculating the moduli of those eigenvalues, providing none of them is equal to unity. If we designate by $W^s(\cdot)$ and $W^u(\cdot)$ the stable and the unstable manifolds, respectively, and by n_s and n_u the number of eigenvalues of $DP(x_0)$ with modulus smaller and greater than one, respectively, we shall have

$$\dim W^s(x_0) = n_s = \dim W^s(\gamma) - 1,$$
$$\dim W^u(x_0) = n_u = \dim W^u(\gamma) - 1.$$

The Poincaré map can thus serve to construct a linear approximation of the structure of the orbits near a closed orbit. Results obtained in this manner are local in the sense that they only hold within a narrow tube around the orbit, but not in the sense that they refer to a short segment of the tube.

Case (ii). Periodically forced systems

A similar procedure can be followed in the case of non-autonomous, periodically forced oscillations. For example, consider the system of differential equations:

$$\dot{x} = f(x,t), \qquad x,t \in \mathbb{R}^n \times \mathbb{R}, \tag{4.6}$$

where f is periodic of period T in t, i.e., $f(x,t) = f(x,t+T)$, $\forall\, x \in \mathbb{R}^n$.

As is well known, system (4.6) can be transformed into an autonomous system of dimension $(n+1)$ by treating time both as the independent variable and a phase space variable. Formally we introduce the function

$$\theta : \mathbb{R}^1 \to S^1$$

$$t \mapsto \theta(t) = \omega t \bmod 2\pi,$$

ω designating the frequency of the forcing term. Equation (4.6) becomes

$$\dot{x} = f(x,\theta),$$
$$\dot{\theta} = \omega, \qquad (x,\theta) \in \mathbb{R}^n \times S^1. \tag{4.7}$$

This representation can be thought of as a cylindrical phase space where the values of θ are restricted to $0 \le \theta \le 2\pi$.

In this case it is possible to define a (global) cross-section

$$\Sigma = \{(x,\theta) \in \mathbb{R}^n \times S^1 \mid \theta = \theta_0 \in (0, 2\pi]\}.$$

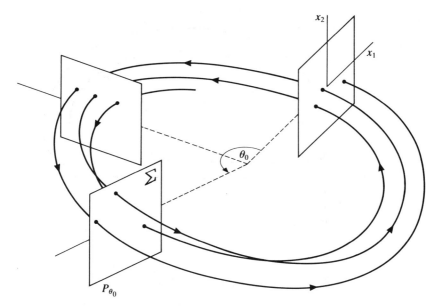

Fig. 4.2. The Poincaré map of a periodically forced dynamical system.

It is evident that all solutions must cross Σ transversally,[6] and that they must intersect it in a recurrent way.[7] This is illustrated in Fig. 4.2.

We can also write a corresponding Poincaré map $P_{\theta_0} : \Sigma^{\theta_0} \to \Sigma^{\theta_0}$ which is defined globally and is given by

$$P(x_0) = \pi \cdot \phi_\tau(x_0, \theta_0), \qquad (4.8)$$

where ϕ_τ is the flow generated by the vector field (4.7); π designates projection onto the first factor; τ is the period of the forcing term and it is the same for all points $x \in \mathbb{R}^n$.

Again, a fixed point of the Poincaré map corresponds to a periodic orbit for the flow, and periodic points of period $k > 1$ for the map correspond to subharmonics of period kT for the flow.

Case (iii). Poincaré maps near a homoclinic orbit

The construction of a Poincaré map in the case of systems possessing ho-

---

[6] Consider that the unit normal to Σ^{θ_0} is $(0, 1)$ and $\langle (f(x, \theta), \omega), (0, 1) \rangle = \omega \neq 0$.

[7] The comment in note 4 applies here. However, in this case, conjugacy and dynamical equivalence are global.

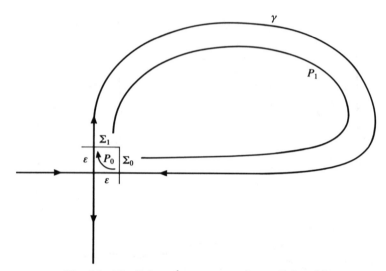

Fig. 4.3. The Poincaré map near a homoclinic orbit.

moclinic (or heteroclinic) orbits is a more complex and delicate operation, and will only be briefly mentioned here.[8]

Let us consider the usual system of differential equations

$$\dot{x} = f(x) \tag{4.9}$$

where $x \in \mathbb{R}^{s+u}$ and $f : U \to \mathbb{R}^{s+u}$ is C^r ($r \geq 2$) on some open set $U \subset \mathbb{R}^{s+u}$.

Suppose that (4.9) has a hyperbolic fixed point at $x = x_0$ and the Jacobian matrix $Df(x_0)$ has s eigenvalues with negative and u eigenvalues with positive real parts. Suppose also that (4.9) has a homoclinic orbit γ connecting x_0 to itself, i.e, there exists a solution $x(t)$ of (4.9) such that $\lim_{t \to +\infty} x(t) = \lim_{t \to -\infty} x(t) = x_0$.

In this case, it is often possible to construct a Poincaré map defined in a neighbourhood of the homoclinic orbit as a composition of two maps, namely $P = P_1 \circ P_0$. The first map, P_0, will be defined near the fixed point; the other, P_1, will be defined away from x_0, but within a narrow tubular neighbourhood of the homoclinic orbit.

A simple two-dimensional case is illustrated in Fig. 4.3. The approximate Poincaré maps will be $P_0 : \Sigma_0 \to \Sigma_1$ and $P_1 : U \subset \Sigma_1 \to \Sigma_0$ where

[8] For a thorough discussion of this question, see Wiggins (1988, pp. 171–334).

the domains Σ_0 and Σ_1 are defined as follows:

$$\Sigma_0 = \{(x, y) \in \mathbb{R}^2 \mid y > 0, \ x = \epsilon > 0\}$$
$$\Sigma_1 = \{(x, y) \in \mathbb{R}^2 \mid y = \epsilon, \ x > 0\}$$

ϵ being a small quantity.

Since, in general, we cannot solve system (4.9) explicitly, the construction of these maps will involve certain local approximations. However, if the neighbourhood around the fixed point and the tube around the homoclinic orbit are chosen small enough, the dynamics of the approximate composite Poincaré map accurately reflect that of the original flow.

An early application of this method can be found in the celebrated study by Šilnikov (1965) of a third-order system of differential equations (see Fig. 4.4). A more recent interesting application to the famous Lorenz model has been provided by Sparrow (1986).

When none of the three 'canonical' cases (periodic orbit; periodically forced oscillations; homo/heteroclinic orbit) occurs, the analytical construction of Poincaré maps associated with flows can sometimes be attempted by means of certain perturbation methods (e.g., averaging) which will not be discussed here.[9]

In many interesting but complicated situations, it may not be possible to derive a discrete-time mapping from the original, continuous-time system in a rigorous manner. In these cases, one may try a drastic simplification, formulating the problem directly in terms of a discrete system, without making recourse to a surface of section and the associated first-return map. This strategy, which was followed, for example, by Hénon with regards to the Lorenz model,[10] has the advantage of reducing computing time (often by a factor of the order of 10^3) and of permitting the use of the available theoretical results on mappings. However, strictly speaking, this amounts to re-modelling the dynamics of the original system, and poses certain methodological questions. If the new (discrete-time) model provides an appropriate description of the problem at hand, one may wonder why it was not chosen to begin with. If, on the contrary, the continuous-time formulation is deemed to be preferable, then the associated first-return map, whenever its construction is possible, is aimed to gain a better understanding of the behaviour of the original flow. The map, therefore, in order to perform this function, cannot

---

[9] For a detailed discussion, see Wiggins (1990, pp. 103–53).
[10] Cf. Hénon (1976).

be formulated independently of the continuous-time system, but must instead be derived from it, in a more or less rigorous manner.

4.3 NUMERICAL POINCARÉ MAPS

When the analytical construction of a Poincaré map is not possible, one can still make recourse to numerical simulation. Thus, one can first integrate the equations of the system, then construct a surface of section, and finally calculate an approximate first-return map, following certain procedures available in the attached software program.

The construction of a Poincaré map by numerical-graphical techniques, however, is a rather delicate procedure which must be performed after careful investigation of the geometric properties of the (apparent) attractor of the system. In particular, one must verify the existence of *recurrence* (i.e., the orbits must repeatedly cross the selected surface) and *transversality* (i.e., the proposed surface must never be tangent to the orbits. Fortunately, however, we know that certain essential quantitative properties of the dynamics of a first-return map (e.g., the Lyapunov spectrum) do not depend in an essential manner on the precise position of the surface of section.

Altogether, experience suggests that, even though the picture of a Poincaré map of an attractor may not always specify its characteristics unambiguously, its study provides very precious clues to the dynamics of the underlying system. This is especially so, when some other independent (and concurrent) information may be obtained, analytically or numerically.

In the simple case in which the surface of section has dimension two, a classification of typical (approximate) Poincaré maps is possible, each type being suggestive of a certain kind of dynamic behaviour. The most common occurrences are illustrated in Fig. 4.5.

We have already observed that a map consisting of *one single point* corresponds to a closed orbit, i.e., in dissipative systems, to a limit cycle. Similarly, a map consisting of a *finite number of points*, visited by the orbits of the system recursively in a predetermined order, suggests *subharmonic oscillations*.

Maps consisting of (an indefinitely large number of) points lying on a *close curve* suggest the presence of *quasiperiodic oscillations*. The latter, it will be remembered (see above, Chapter 2), take place when the dynamical system is a combination of two or more oscillators, with frequencies incommensurate with respect to one another.

When the (indefinitely large number of) strobe points lie on an *open*

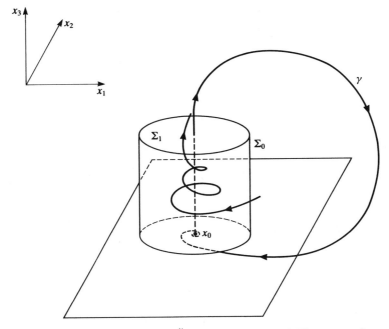

Fig. 4.4. The Poincaré map of the 'Šilnikov phenomenon'. The two surfaces are
defined thus:

$$\Sigma_0 = \{(x_1, x_2, x_3) | x_1^2 + x_2^2 = r_0^2, \text{ and } 0 < x_3 < \bar{x}\};$$

$$\Sigma_1 = \{(x_1, x_2, x_3) | x_1^2 + x_2^2 < r_0^2, \text{ and } x_3 = \bar{x} > 0\}.$$

curve, a low-dimensional *strange attractor* of a single sheet type may be
present. The sheet presumably consists of an infinite number of closely
spaced 'leaves'. The presence of rapid contraction of volumes, associated
with strong dissipation of the system, makes the Poincaré map appear as
if it were one-dimensional. In practice, this effect is observed when the
rate of volume contraction is of order n^2, n being the dimension of the
phase space.

In this situation, one is encouraged to model the system as a one-
dimensional map whose behaviour will presumably reflect that of the
original system, following the so-called '1D map approach'.[11]

A *fractal collection of points* (like the one characterizing the attractor

---

[11] See, for example, our discussion of Chapter 11 below, where we provide a one-
dimensional representation of a continuous-time model whose rate of contraction is
actually n^2.

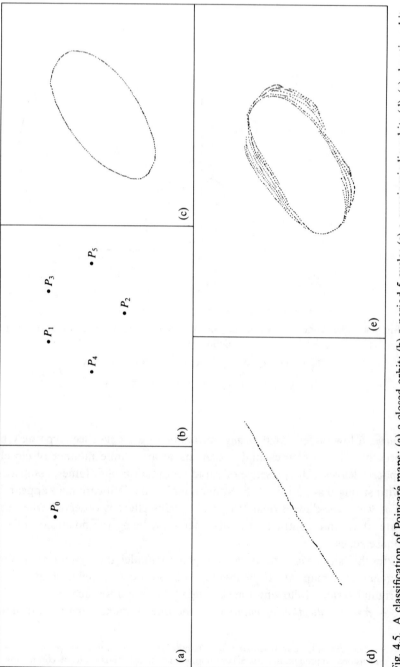

Fig. 4.5. A classification of Poincaré maps: (a) a closed orbit; (b) a period 5 cycle; (c) a quasiperiodic orbit; (d)–(e) chaotic orbits.

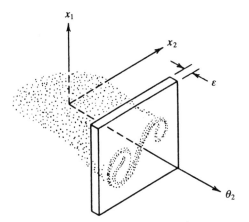

Fig. 4.6. A double Poincaré map.

of our own model of Chapter 12) suggests the presence of a *strange attractor*. If there is indeed such an attractor, this will presumably have a fractal dimension between two and three, and will be geometrically the product of a two-dimensional surface and a Cantor set.

A *fuzzy collection of points* suggests a more complicated situation. If the system is noise-free, there are three basic possibilities, namely: (i) a *strange attractor* 'living' in a phase space of dimension greater than three; (ii) *quasiperiodic oscillations* with more than two incommensurate frequencies; (iii) *a strange attractor* in a phase space with three dimensions, but with such a weak dissipation that its structure is not reflected properly by the surface of section.

Case (iii) may sometimes be verified by calculating the Lyapunov characteristic exponents of the system (see Chapter 6) and/or one of the various estimates of the fractal dimension of the attractor (see Chapter 7).

One may also try to ascertain the presence of higher dimensional strange attractors by means of the so-called *multiple Poincaré map*. The simple case of a double Poincaré map is illustrated in Fig. 4.6 and can be dealt with as follows.[12] Suppose we have a system of two differential equations with a forcing term characterized by two distinct, incommensurate frequencies, namely:

$$\dot{x} = f(x, \omega_1 t, \omega_2 t), \tag{4.10}$$

[12] Cf. Moon (1987, pp. 142–47).

where $(x, t) \in \mathbb{R}^2 \times \mathbb{R}$; $f: \mathbb{R}^2 \times \mathbb{R} \to \mathbb{R}^2$ is periodic of period 2π in the last two arguments, and ω_1/ω_2 is irrational. We assume that $\omega_1 > \omega_2$.

As we know from the previous discussion, system (4.10) can be re-written in an autonomous form, thus:

$$\dot{x} = f(x, \theta_1, \theta_2)$$
$$\dot{\theta}_1 = \omega_1 \qquad\qquad (4.11)$$
$$\dot{\theta}_2 = \omega_2$$

where $(x, \theta_1, \theta_2) \in \mathbb{R}^2 \times S^1 \times S^1$.

We can then define a first cross-section

$$\Sigma_1 = \{(x, \theta_1, \theta_2) \in \mathbb{R}^2 \times S^1 \times S^1 | \theta_1 = \bar{\theta}_1 \in (0, 2\pi]\},$$

and construct the Poincaré map associated with it, i.e.,

$$P_{\bar{\theta}_1} : \Sigma_1 \to \Sigma_1.$$

If the latter results in a fuzzy set of points on Σ_1, we define a second set

$$\Sigma_2 = \{(x, \theta_1, \theta_2) \in \mathbb{R}^2 \times S^1 \times S^1 | \; \theta_2 \in (0, 2\pi], |\theta_2 - \bar{\theta}_2| \le \epsilon\},$$

which can be interpreted as a finite-width 'slice' of a cross-section associated with the forcing term with frequency ω_2.

We now store only those points at which trajectories intersect Σ_1 and Σ_2 simultaneously (i.e., we consider the set $\Sigma_1 \cap \Sigma_2$).

Applications of this idea, sometimes called a 'Lorenz cross-section', to four-dimensional continuous systems[13] have revealed the fractal character of an attractor which did not appear in a three-dimensional Poincaré map, nor in its projection onto a plane in two of the variables.

The technique can be generalized to higher dimensional maps. However, there are practical limits to it. On the one hand, the width of the 'slice', 2ϵ, must be much smaller than 2π. On the other hand, unless the frequency of the relevant forcing term is very large, the probability of three or more simultaneous intersections may be extremely small, making the experiment impracticable.

Remark 4.1. For low-dimensional systems, the Poincaré map procedure may be used to infer the fractal dimension of the underlying continuous system. For example, if $0 < d < 2$ is the dimension of the attractor of the map, evaluated by one of the various techniques discussed below (see Chapter 7), then

$$D = d + 1$$

---

[13] Lorenz (1984, pp. 90–104) and Moon and Holmes (1985, pp. 157–60).

is a good estimate of dimension of the 'complete' attractor pertaining to the continuous system.

Remark 4.2. We have so far considered the problem of deriving a map from a flow. Sometimes, however, it is important to know whether and under what conditions, a map can be regarded as the first-return map of a flow. First of all, if the map is continuous but not invertible, it can never be conceived of as a Poincaré map of a flow, simply because flows have unique orbits through each point. However, in certain cases, it is possible to relate a two-to-one, non-invertible map to *one coordinate* of a two, or higher dimensional surface of section, rather than to the entire surface.[14]

An application of this idea is in fact the '1D map approach', which we have already mentioned, and which has proved a very useful (though approximate) tool of analysis.

Remark 4.3. There exists an interesting relationship between the entropy computed with respect to a continuous-time dynamical system, and that of a Poincaré map derived from it, known as *Abramov formula* (cf. Eckmann and Ruelle, 1985, p. 637). If ρ is a measure invariant under the action of a flow, and σ is a measure, corresponding to ρ, invariant under a Poincaré map associated to that flow (i.e., σ is the density of intersections of orbits with a surface of section Σ transverse to the flow), then we have

$$h(\rho) = \frac{h(\sigma)}{\langle \tau \rangle_\sigma},$$

where $h(\rho)$ denotes (Kolmogorov–Sinai) entropy as defined in Chapter 2 above and $\langle \tau \rangle_\sigma$ is the average time between two intersections, calculated with respect to σ. Notice that, since $\langle \tau \rangle_\sigma$ is essentially positive, $h(\sigma) > 0 \iff h(\rho) > 0$, so that chaoticity of the Poincaré map implies that of the associated flow (and vice versa).

---

[14] Cf. Jackson (1989, vol. I, pp. 206–9).

5

Spectral analysis

5.1 THE FOURIER TRANSFORM

In economics, the analysis of the time evolution of variables is largely performed in terms of dynamical models whose solutions consist of functions of the independent variable, time. The properties of those functions are often studied through their projections onto the phase space. This is known as the *time-domain* approach.

However, the understanding of the behaviour of economic systems can be significantly enhanced by extending investigation of them to the *frequency domain*, by means of so-called spectral analysis. There exists a very substantial literature on this subject,[1] and we shall limit ourselves here to a survey of some basic concepts and a few considerations concerning the application of spectral analysis to the study of chaotic dynamics.

A basic result in this field is given by the Fourier theorem, which, in a nutshell,[2] states that any p-periodic function $f(t)$ can be represented as a convergent trigonometric series, i.e.,[3]

$$f_p = \frac{A_0}{2} + \sum_{k=1}^{+\infty} \left(A_n \cos \frac{2\pi kt}{p} + B_n \sin \frac{2\pi kt}{p} \right),$$

or, in a more convenient, complex form

$$f_p = \sum_{k=-\infty}^{+\infty} a_k e^{2\pi ikt/p} \tag{5.1}$$

---

[1] See, for example, Priestley (1981), and, for application-oriented readers, Press *et al.* (1986, pp. 381–453).

[2] A more detailed discussion can be found, for example, in Sirovich (1988, pp. 139–52).

[3] Equality only holds if $f(t)$ is continuous. If $f(t)$ is not continuous, but piece-wise differentiable, then the RHS of (5.1) will be equal to the arithmetic mean of the limits of the function on both sides of the discontinuity.

where $i = \sqrt{-1}$, and[4]

$$f_p = f(t), \qquad \frac{-p}{2} < t < \frac{p}{2}.$$

The complex coefficients a_k will be determined by

$$a_k = \frac{1}{p} \int_{-p/2}^{p/2} f_p(t)e^{-2\pi ikt/p}dt. \tag{5.2}$$

Roughly speaking, the term a_k represents the contribution of the kth frequency component to series (5.1).

Since we are mainly concerned here with aperiodic processes, we need an extension of the Fourier theorem to nonperiodic functions. The idea is to treat nonperiodic functions as functions of infinite period by taking the limit $p \to \infty$. Since the coefficients a_k vanish under this limit, we shall instead consider the product a_kp which exists under the limit. Now, let us define

$$\Delta\omega = \frac{1}{p},$$

where Δ denotes the interval between two consecutive samples of ω, and

$$a_kp = F(k\Delta\omega),$$

where $F(k\Delta\omega)$ can be viewed as a sampled form of a continuous function $F(\omega)$. We can now rewrite (5.2) and (5.1), respectively, as

$$F(k\Delta\omega) = \int_{-p/2}^{p/2} f_p(t)e^{-2\pi it(k\Delta\omega)}dt$$

$$f_p = \Delta\omega \sum_{k=-\infty}^{+\infty} F(k\Delta\omega)e^{2\pi it(k\Delta\omega)},$$

whence, taking the limit $p \to \infty$ or $\Delta\omega \to 0$, we obtain

$$F(\omega) = \int_{-\infty}^{+\infty} e^{-2\pi i\omega t}f(t)dt, \tag{5.3}$$

and

$$f(t) = \int_{-\infty}^{+\infty} e^{2\pi i\omega t}F(\omega)d\omega. \tag{5.4}$$

Equations (5.3) and (5.4) – the so-called 'Fourier pair' – can be thought of as two different representations of the same process. Equation (5.4), on the one hand, describes the process in the time domain, giving us the values of a certain quantity as a function of time, t. Equation (5.3) (known as the Fourier transform of $f(t)$), on the other, specifies the same

---

[4] The restriction of $f(t)$ to the interval $(-p/2, p/2)$ is not important since an elementary change of variables can always reduce an arbitrary interval (a, b) to $(-p/2, p/2)$.

process in the frequency domain by giving its amplitude and phase ($F(\omega)$ is, in general, a complex quantity) as a function of frequency ω.[5]

5.2 POWER SPECTRAL DENSITY

The *power spectrum* $P(\omega)$ (or power spectral density, PSD) of a scalar signal $f(t)$ is defined as the square of (the modulus of) its Fourier coefficient, i.e., $|F(\omega)|^2$, and it measures the energy per unit of time, or the power of the signal, as a function of the frequency ω.[6]

Often one is interested in knowing how much power is contained in a signal over a certain frequency interval $(\omega, \omega + d\omega)$ and does not wish to distinguish between positive and negative frequencies.

Then one defines the *one-sided* PSD as

$$P(\omega) = |F(\omega)|^2 + |F(-\omega)|^2, \qquad 0 \le f < \infty.$$

If $f(t)$ is a real function, as is normally the case in economic applications, then

$$|F(\omega)| = |F(-\omega)|,$$

and therefore

$$P(\omega) = 2|F(\omega)|^2,$$

(although the factor 2 is often omitted).

A very interesting relation between time and frequency representations of a signal can be shown by means of the idea of *convolution*.[7] The convolution of two functions $f(t)$ and $g(t)$, denoted by $f * g$, is defined as

$$f * g = \int_{-\infty}^{+\infty} f(t)g(t - \tau)d\tau,$$

and it can be shown that $f * g = g * f$.

---

[5] A condition for the Fourier transform to exist is that the function $f(t)$ be absolutely integrable over the real line, i.e., that we have

$$\int_{-\infty}^{+\infty} |f(t)|dt < \infty.$$

This implies that $\lim_{t \to \infty} f(t) = 0$ and may sometimes be a too severe constraint. Fortunately, the condition can be loosened somewhat, and convergence can be proved to exist when $f(t)$ is *Hölder-continuous*. This means that at each point t there exists a constant K such that

$$|f(t + \delta) - f(t)| < K|\delta|^\alpha, \qquad \alpha > 0.$$

[6] To be more precise, the quantity $|F(\omega)|^2$ must be calculated by averaging over small intervals of ω, so as to avoid fluctuations.

[7] See Sirovich (1988, pp. 260–93).

It turns out that the Fourier transform of the convolution of two functions is just the product of the Fourier transforms of the individual functions, i.e., we have:

$$f * g \Longleftrightarrow F(\omega)G(\omega), \qquad (5.5)$$

where the symbol $\Longleftrightarrow$ is used here to designate the transform pairs.

Consider that the *correlation* of two functions, denoted by $\mathrm{Corr}(f, g)$, is defined as

$$\mathrm{Corr}(f, g) = \int_{-\infty}^{+\infty} f(t + \tau)g(\tau)d\tau.$$

The correlation is a function of time and, as we shall see later on (see Chapters 11 and 13), it is of special interest for discussion of lags in economic models.

From (5.5) and the fact that, for real functions, $G(-\omega) = G^*(\omega)$ (where $G^*(\omega)$ denotes the complex conjugate of $G(\omega)$), it follows that

$$\mathrm{Corr}(f, f) \Longleftrightarrow F(\omega)F^*(\omega) = |F(\omega)|^2. \qquad (5.6)$$

This transform pair, known as the *Wiener–Khinchin theorem*, implies that the auto-correlation function and the PSD of a function of time $f(t)$ contain the same information on f.

Another interesting result, known as *Parseval's theorem* states that the total power of a function of time is the same in the time domain as in the frequency domain, i.e., that

$$\int_{-\infty}^{+\infty} |f(t)|^2 dt = \int_{-\infty}^{+\infty} |F(\omega)|^2 d\omega.$$

Thus, power spectral analysis may be viewed as a method for determining the fraction of the total variance to be assigned to frequencies over a certain interval.

5.3 APPLICATIONS OF SPECTRAL ANALYSIS

Since in Part II of this book we make repeated use of spectral analysis in the discussion of different economic models, this section will include only a few broad comments.

Spectral analysis can be effectively employed in the investigation of dynamical systems.[8] It can provide us with a representation of the output of a system from which we can often derive precious (qualitative as well as quantitative) indications concerning its mode of behaviour. It can also

---

[8] In this work we deal almost entirely with deterministic systems. For a discussion of applications of spectral analysis to stochastic processes, besides the quoted works, see Granger and Newbold (1977, pp. 43–71).

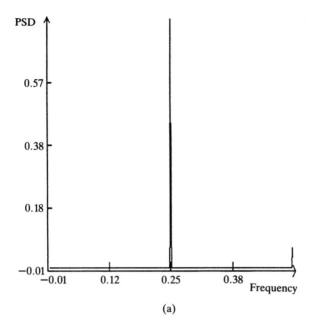

(a)

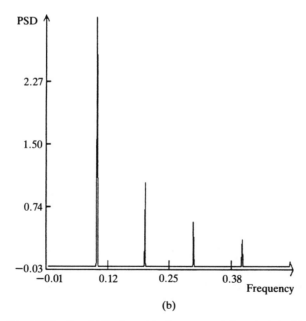

(b)

Fig. 5.1. PSD of periodic dynamical systems: (a) period 4; (b) period 10.

be usefully applied to the analysis of experimental signals, i.e., actual time
series, although in this case certain serious problems may arise, owing to
the inevitable presence of noise.

By looking at the structure of PSD, we can draw some interesting conclusions concerning periodicity or quasiperiodicity of dynamical systems.
The simplest case is that of *periodic* systems. In this case, the spectrum
consists of a set of Dirac delta functions located at the fundamental
frequency and perhaps at its subharmonics. In practice, a numerical
spectral density function reveals the presence of a δ function by means
of 'peaks' neatly marked and regularly spaced, but of finite height.

A very simple example of a discrete-time periodic model is the following:

$$x_{t+1} = x_t^P = x_t \bmod P + 1,$$

where P is the parameter controlling the length of the period and as
usual we have

$$x \bmod P = \begin{cases} x, & \text{for } x < P \\ x - P, & \text{for } x \ge P. \end{cases}$$

The time series x_t^P thus generated is of the type called 'saw-tooth'. If we
had an infinitely long string of data we would obtain a Fourier series
consisting of an infinite number of sinusoidal terms with frequencies
which are multiples of the fundamental one, $1/P$. Since this is an odd
function, its Fourier series would contain only sine functions and, putting
for example, $P = 10$, we would have

$$x(t) = 5 - 2\sin\left(\frac{\pi t}{5}\right) - \sin\left(\frac{2\pi t}{5}\right) - \frac{2}{3}\sin\left(\frac{3\pi t}{5}\right) - 2\sin\left(\frac{4\pi t}{5}\right) - \cdots \quad (5.7)$$

The function x_t^P can be viewed as a discrete-time version of (5.7), and
therefore it has only a finite number of harmonics with a maximum
frequency given by the so-called 'Nyquist critical frequency', of which
more in a while.

We can verify the argument above by examining Figs. 5.1a and 5.1b,
showing the PSD of x_t^P with $P = 4$ and with $P = 10$, respectively. The
reader will notice that in the second case we have significant peaks at
four frequencies which are multiples of $1/10$. In the first case, instead,
only one frequency different from zero appears in the PSD (at $\omega = 1/4$).

We can see that, in this simple case, the periodic character of a model
is clearly detectable and its spectrum consists entirely of (approximate)
δ functions. When we meet a spectrum of the shape just described, it is
safe to exclude more complex behaviour. However, we should hasten to
add that great care must be given to the choice of data to be analysed by

means of PSD. For example, inclusion of transients could generate peaks similar to those produced by a periodic model, together with various subharmonics of relatively low amplitudes, which, if one is interested in asymptotic behaviour, should be ignored.

A quasiperiodic model involving fundamental frequencies $\omega_1, \ldots, \omega_n$ will have a power spectrum consisting of δ functions corresponding to those frequencies and to all linear combinations of them with integer coefficients.

A simple example is given by the system

$$\theta_{n+1} = \theta_n + \Omega \bmod 1, \tag{5.8}$$

which can be thought of as the Poincaré map of a flow on a T^2 torus, characterized by two frequencies ω_1 and ω_2, such that $\omega_1/\omega_2 = \Omega$. If Ω is irrational, the motion will be quasiperiodic, i.e., the orbits will cover the whole torus and correspondingly the iterations of the map will cover the whole circle, without ever becoming periodic. Fig. 5.2 illustrates the power spectrum of a typical trajectory of (5.8) for $\Omega = \sqrt{2}$. Notice, however, that, because of finite resolution, the numerically computed spectrum of a quasiperiodic trajectory cannot be rigorously distinguished from that of a periodic one characterized by a fundamental frequency and its harmonics.

In Figs. 5.3a and 5.3b we can see, respectively, a time series of a white-noise process, generated through appropriate numerical algorithms, and its spectral density. The latter indicates that there exist non-zero components for all frequencies. Theoretically, the spectrum amplitude should be constant, i.e., independent of the frequency.

On purely *a priori* grounds, one would expect that the 'stochastic behaviour of a deterministic dynamical system' should be signalled by a PSD qualitatively similar to that of a truly random time series. Indeed, a *broad-band power spectrum* is a typical symptom of chaotic behaviour. The presence of sharp peaks, however, does not necessarily exclude chaos. Certain embedded periodicities may be present in otherwise chaotic behaviour. This type of chaos, characterized by the simultaneous presence in the spectrum of broad bands and peaks, is sometimes called 'non-mixing' (cf. Oono and Osikawa, 1980). This is the case, for example, of the well-known Rössler attractor. The power spectra of systems characterized by mixing[9] and non-mixing chaos are discussed below in Chapters 11 and 13.

[9] The notion of *mixing* can be made more precise in terms of measure theory which we have discussed concisely in section 2.8 above. (See Halmos, 1956, pp. 36–7). Let T be a

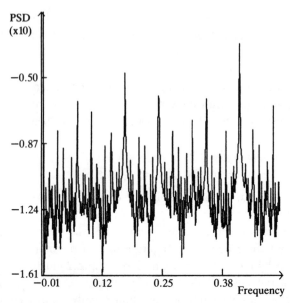

Fig. 5.2. PSD of a quasiperiodic system.

For many chaotic systems, it has been observed (see Shaw, 1980, p. 107) that in addition to a certain noise floor, there exists so-called 'flicker noise' or 'excess noise'. This is sometimes defined as '$1/f^\alpha$ noise' because the low frequency components dominate over the high frequency ones approximately according to an exponential function.

We do not yet have a rigorous explanation of this occurrence. It seems, however, that it is a common feature of a vast class of chaotic systems. The investigation of flicker noise in the Lorenz attractor explains the predominance of low-frequency components by the presence of unstable fixed points. When the trajectories pass near those points – so the

transformation preserving the measure μ of the underlying space Ω, and F, G any pair of measurable subsets of Ω. Then we say that T is *mixing* if $\lim_{k\to\infty} \mu(T^{-k}F \cap G) = \mu(F)\mu(G)$. A physical, and somewhat jocular interpretation of this is the following. A cocktail shaker contains a fixed amount of liquid consisting of, say, 90% Coca-Cola and 10% rum, and the transformation T is a particular way of shaking the liquid. If G is the region originally occupied by the rum, then for any part F of the shaker, the relative amount of rum in F, after k shakes, is given by $\mu(T^{-k} \cap G)/\mu(G)$. If T is mixing (i.e., if the barman is good at making cocktails), then after a sufficiently large number of shakes, every part F of the shaker will contain approximately 10% of the rum. In a system with mixing, it is essentially impossible to distinguish points that were or were not in a certain set F a sufficiently long time earlier. Thus, mixing can be taken to characterize chaos. Notice that mixing implies ergodicity but not vice versa.

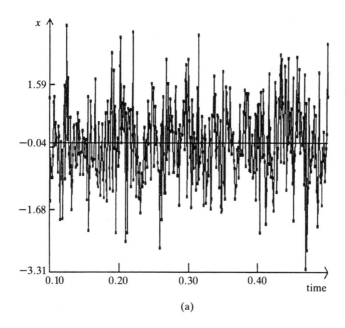

(a)

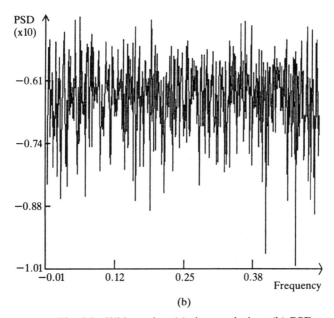

(b)

Fig. 5.3. White noise: (a) time evolution; (b) PSD.

argument runs – they are drawn out of the chaotic zone of the attractor, and they return to it only after a more or less long time lag. A more detailed discussion of this phenomenon, which is called 'intermittency', can be found in Chapter 9.[10]

In conclusion, we can say that power spectrum analysis is quite good at telling periodic from aperiodic signals and identifying periodicities embedded in chaotic bands. However, this method has serious limitations. In practice, it can be difficult to distinguish, by means of PSA, between high-order quasiperiodic signals and chaotic ones. Moreover, it is often important to distinguish between low and high-dimensional chaotic attractors; but for this purpose, PSA is not particularly useful, because a broad-band power spectrum is common to the two cases. Finally, since, the 'weights' of the PSD are given by the squares of the moduli of the Fourier coefficients, some of the information contained in the original signal (which could be precious in the investigation of chaotic behaviour), is lost.

5.4 THE ESTIMATE OF PSD. ALIASING

When estimating PSD, we commonly deal with sampled values of a variable, i.e., values recorded at certain intervals of time. This is true both when we integrate ordinary differential equations numerically, and when we investigate real time series (experimental signals).

If we denote by Δ the time interval between two consecutive samples, we can describe the sequence of data as

$$z_k = z(k\Delta), \qquad k = 1, \ldots$$

For any given interval Δ, we can define a frequency

$$\omega_c = \frac{1}{2\Delta}$$

which is called the *Nyquist critical frequency*.

The Nyquist frequency is important since it is the maximum frequency detectable from series with a sampling rate equal to Δ. This can be expressed by saying that a sinusoid of frequency ω_n can be sampled only at its maximum or minimum points. How does this limitation affect the analysis of dynamic processes?

In practice, one has sometimes to deal with functions with a bandwidth limited by ω_c (i.e., $F_x(\omega) = 0$ when $|\omega| > \omega_c$), and, in this case, a known result called 'the sampling theorem' guarantees that, if we had an infinite

---

[10] On the relation between $1/f$ noise and intermittency, see Schuster (1989, pp. 91–8).

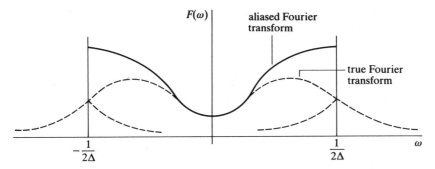

Fig. 5.4. Aliasing.

number of samples z_k, the function $z(t)$ would be completely determined by them.[11] However, if we are dealing with a process which is not bandwidth-limited, the spectral representation of the function under scrutiny may be seriously distorted by a phenomenon called the 'aliasing effect'. The effect of 'aliasing' is that any frequency outside the critical interval $[-\omega_c, \omega_c]$, is projected onto it because of the sampling. This causes distortions especially at high frequencies (see Fig. 5.4).

The only way to reduce the negative consequences of the aliasing effect is to sample the process at sufficiently short intervals, so as to obtain two points for each cycle even at the highest frequency present. Clearly, if the latter is not known, the only strategy is to sample as often as possible.

5.5 ESTIMATING PSD WITH FINITE SAMPLES

In practice, one has to estimate the Fourier transform from a finite number of values generated by numerical integration of a system of o.d.e. by the iteration of a map, or obtained directly from a real time series.

Suppose we have N consecutive sampled values

$$z_k \equiv z(t_k), \qquad t_k = k\Delta, \qquad k = 0, 1, 2, \ldots, N-1,$$

Δ being the sampling interval.

Intuitively, since we have N samples, we cannot have more than N independent estimates of the transform. Therefore, rather than estimating

---

[11] We could also say that, a function with a limited bandwidth carries an informative content which is, in a sense, infinitely smaller than that of a general continuous function and this is the reason why a complete knowledge of the former can, in principle, be obtained through discretely sampled data.

it over the whole set of frequencies $[-\omega_c, \omega_c]$, we can limit ourselves to the discrete frequencies

$$\omega_n \equiv \frac{n}{N\Delta}, \qquad n = -N/2, \ldots, N/2.$$

(Notice, that there are $N+1$ values of ω_n, but the first and the last are not independent.) We must now approximate the Fourier transform as

$$F(\omega_n) = \int_{-\infty}^{+\infty} f(t) e^{2\pi i \omega_n t} dt \approx \sum_{k=0}^{N-1} z_k e^{2\pi i \omega_n t_k} \Delta = \Delta \sum_{k=0}^{N-1} z_k e^{2\pi i k n / N}.$$

The last summation above, i.e.,

$$Z_n = \sum_{k=0}^{N-1} z_k e^{2\pi i k n / N}, \tag{5.9}$$

is known as the discrete Fourier transform (henceforth DFT) of the function $z(t)$.

From (5.9) we can obtain a rough estimator of PSD called a *periodogram* and defined as follows:

$$P(0) = P(\omega_0) = \frac{1}{N^2} |Z_0|^2,$$

$$P(\omega_j) = \frac{1}{N^2} \left[|Z_j|^2 + |Z_{N-j}|^2 \right], \quad j = 1, 2, \ldots, N/2 - 1, \tag{5.10}$$

$$P(\omega_c) = P(\omega_{N/2}) = \frac{1}{N^2} |Z_{N/2}|^2,$$

where ω_j is only defined for zero or positive frequencies, i.e.:

$$\omega_j \equiv \frac{j}{N\Delta} = 2\omega_c \frac{j}{N}, \qquad j = 0, 1, \ldots, N/2,$$

and the terms $P(\omega_s)$ indicate the power at $N/2 + 1$ frequencies. We said that the periodogram is a rough estimate of PSD since it is not a correct estimator. This can be deduced intuitively on the basis of the observation that we are deriving estimates of frequencies which are separated by a constant lag. That is to say, we can estimate components at frequencies ω_i and ω_{i+1}, but we can say nothing of the PSD concerning the intermediate frequencies. The information contained in the intermediate components is not lost, however, but it affects the others. The value of a component of frequency ω_i is in fact a sort of average of all the values of the components which have not been estimated, and which have frequencies included in an interval I_i centred in ω_i. This phenomenon is called 'leakage'.

The length of the interval I_i varies according to the type of periodogram employed and we shall not discuss it here. Suffice it to say that in many cases its length is fairly large and that therefore the distortion of the periodogram may be large (for further details, see Press *et al.*, ch. 12.7).

To overcome this problem, one can employ the technique called 'data windowing'. The most frequently adopted types of lag windows are the following:

Parzen window

$$\lambda_\tau = 1 - \left| \frac{\tau - \frac{1}{2}(N-1)}{\frac{1}{2}(N-1)} \right|$$

Welch window

$$\lambda_\tau = 1 - \left(\frac{\tau - \frac{1}{2}(N-1)}{\frac{1}{2}(N-1)} \right)^2$$

Hanning window

$$\lambda_\tau = \frac{1}{2} \left[1 - \cos\left(\frac{2\pi j}{N-1} \right) \right],$$

among which, the most popular is perhaps the Parzen window.

A further problem concerning the estimate of PSD is that of the inconsistency of the periodogram. The periodogram, being an estimator and therefore a random variable, has its own variance. Were the estimator consistent, its variance would tend to zero as the number of observations tended to infinity. Consistency implies that the estimator becomes progressively more accurate as we acquire additional information. The inconsistency of the periodogram can be explained as follows. Suppose we increase the number of data concerning the sampled function, keeping unchanged the sampling interval. The Nyquist interval will not change, but we shall have a larger number of frequencies which are observable within it and will absorb the increase of information. If, on the contrary, we reduce the sampling interval while keeping constant the number of estimates per unit of time, the additional information will result in an increased Nyquist interval. In neither case will the variance of the estimated PSD be reduced.

Finally, there exist certain techniques capable of reducing the variance of the estimate. One of these is so-called 'overlapping'. This consists of a subdivision of a set of sampled data (a time series), into K subintervals of length $2M$[12] for each of which one calculates the Fourier transform. Then, one averages these K transforms in order to obtain a more accurate estimate of PSD. In this way, one reduces the variance by a factor K, thereby increasing the precision of the estimate.

---

[12] For this purpose, it is required that the number of sampled data be a power of 2.

6

Lyapunov characteristic exponents

6.1 THE THEORY OF LCEs

In the introductory discussion of Chapter 2, we have already mentioned that sensitive dependence on initial conditions is the distinctive feature of chaotic dynamical systems. The fact that, owing to SDIC, nearby trajectories become exponentially separated in finite time under the action of a flow (or a map), makes the evolution of those systems very complex and essentially unpredictable, except perhaps in the short run. It is now time to discuss in some detail SDIC and the method to detect its presence in given dynamical systems, as well as in experimental signals.

The most important tool for diagnosing the presence of SDIC in a dynamic system is provided by *Lyapunov characteristic exponents* (LCEs). These are named after the Russian mathematician Lyapunov who started this line of investigation of dynamical systems at the turn of the century (see Lyapunov, 1907). The theory of LCEs in its modern form was first formulated in the late sixties by Oseledec (1968) and underwent substantial development in the following years.[1]

LCEs are important because they allow one to give a precise quantitative definition of SDIC (and thereby of chaoticity), and they are numerically computable.

The discussion that follows will be addressed to continuous-time systems, but all the results just mentioned can be easily applied to the case in which t is an integer, i.e., to discrete-time systems. For this purpose, remember that, as was mentioned in Chapter 2, we can always discretize a continuous-time dynamical system by taking a time-one map derived from its solution. All the quantities that interest us here (e.g., LCEs,

---

[1] For an excellent introductory discussion of the main results and a basic bibliography, see Benettin and Galgani (1979) and Benettin *et al.* (1980) from which we have drawn heavily; more recently, see Wolf *et al.* (1985), which also contains the computer programs for calculating LCEs on which the ones in this work are based.

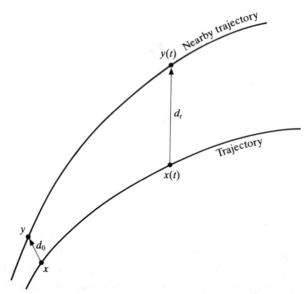

Fig. 6.1. Two nearby trajectories separating as time goes by.

entropy, fractal dimensions, etc.) will remain essentially unchanged under the transformation. Later in this chapter, we shall have something more to say on the relation between the LCEs of a flow and those of the related Poincaré map.[2]

Intuitively, we can describe *LCEs of order one* (i.e., LCEs of a vector) as follows. Consider two arbitrarily close points x and y on the phase space M and take their distance, d_0. Apply now a flow ϕ_t to these points and, after a time t, take again their distance, d_t. Broadly speaking, the ratio between these two distances can be expressed by $e^{\chi t}$. Suppose now that, as $t \to \infty$, χ converges to a limit. This limit is called the LCE. Clearly if $\chi > 0$, we can say that, under the action of the flow, nearby orbits diverge exponentially, i.e., there is sensitive dependence on initial conditions.

The definition of LCEs can be formulated more precisely as follows. Let $\phi_t: M \to \mathbb{R}^n$ be a flow defined by a system of ordinary differential equations

$$\dot{x} = f(x), \qquad x \in \mathbb{R}^n. \tag{6.1}$$

Let $M \subset \mathbb{R}^n$ be the phase space of system (6.1) and $x, y \in M$, and let

[2] Cf. also Ruelle (1989, p. 51).

μ be a probability measure on M which is preserved by ϕ_t.[3] Consider now a point $x \in M$ and a nearby point y. The rate of change of their distance, as measured, for example, by the Euclidean norm $\| \cdot \|$, will evolve under the action of the flow ϕ_t according to the ratio

$$\|\phi_t(y) - \phi_t(x)\|/\|y - x\|. \tag{6.2}$$

The limit of (6.2) $y \to x$ along a given curve γ can be expressed by

$$\|D\phi_t(x)w\|/\|w\|,$$

where $D\phi_t(x)$ is the derivative with respect to x of ϕ_t at x and w is a vector of T_xM (the tangent space to M at x) tangent to γ. Correspondingly, the evolution in time of the tangent vector w will be governed by the so-called variational equation, i.e.,

$$\dot{w} = A(x(t))w, \tag{6.3}$$

where $A \equiv Df(x)$ is the (time-varying) matrix of partial derivatives.

We can now define the *LCE of the vector* w (which in general will depend on x) as follows:

$$\lim_{t \to \infty} \frac{1}{t} \ln \frac{\|D\phi_t(x)w\|}{\|w\|} = \chi(x, w). \tag{6.4}$$

The basic result in this field is the proof that, under rather mild conditions on M and $\phi_t(x)$,[4] the limit (6.4) actually exists, for μ-almost all points $x \in M$ (Oseledec, 1968).

LCEs are most easily understood in the simple case of a linear system

$$\dot{x} = Ax,$$

where A is a constant $(n \times n)$ matrix. In this case, we have:

$$\phi_t(x) = e^{At}x,$$

and

$$D\phi_t(x) = e^{At}.$$

A simple relation can now be established between the eigenvalues of A and LCEs. If we put $B \equiv e^{At}$, we can write

$$Bu_i = \lambda_i u_i \quad i = 1, \ldots, n,$$

where u_i and λ_i are, respectively, the eigenvectors and the eigenvalues of B, and n its dimension. For simplicity's sake, we assume that B has n distinct eigenvalues and we arrange them in decreasing order $|\lambda_1| > |\lambda_2| > \cdots > |\lambda_n|$. If w is an arbitrary vector, in general we can

---

[3] On the concept of invariant measure, see the discussion in Chapter 2.

[4] For example, it is required that the manifold M and the flow ϕ_t be of class $\mathbf{C}^1$.

write

$$w = \sum_{i=1}^{n} c_i u_i, \quad c_1 \neq 0.$$

In the limit as $t \to \infty$, the first term of the sum above dominates the others and we have

$$\frac{\|D\phi_t(x)w\|}{\|w\|} \to \frac{\|Bu_1\|}{\|u_1\|} = |\lambda_1|.$$

Hence,

$$\chi(x,w) = \chi_1 = \frac{1}{t}\ln|\lambda_1|.$$

If we denote by σ an eigenvalue of A, we have $\lambda = e^{\sigma t}$ and since $|e^{\sigma}| = e^{\text{Re}\,\sigma}$, we can conclude that

$$\chi_1 = \text{Re}\,\sigma_1,$$

where χ_1 is the largest LCE of order one and σ_1 is the eigenvalue of A with the largest real part.[5]

All the other LCEs in decreasing order can be determined simply by choosing the vector w appropriately. For example, the second LCE, χ_2, can be determined by putting $c_1 = 0$, and $c_j \neq 0$, for $j = 2, \ldots n$, and all the others can be found in an exactly analogous manner.

A slightly more complicated example is given by a system characterized by a periodic orbit γ. In this case, if τ is the period of the orbit, $D\phi_\tau(x)$ is a linear mapping of the tangent space $T_x M$ onto itself, which, for any given τ, can be written as a constant matrix $e^{R\tau}$, where R is a constant $(n \times n)$ matrix. Suppose now that $e^{R\tau}$ has n distinct eigenvalues $\lambda_i (i = 1, \cdots, n)$, arranged in decreasing order as follows:

$$|\lambda_1| > |\lambda_2| > \cdots > |\lambda_n|,$$

and n corresponding eigenvectors $u_1, u_2, \cdots, u_n$. The eigenvalues λ_i are also known as *Floquet multipliers* of the orbit γ.

If the vector w is again chosen at random, in the limit, as $t \to \infty$, we have

$$\chi_1(x,w) = \frac{1}{\tau}\ln|\lambda_1| = \text{Re}\,\sigma_1,$$

where σ_1 now denotes the dominant eigenvalue of the matrix R. Notice that one of the Floquet multipliers (that corresponding to displacements along the orbit γ) is always equal to unity. Therefore, one of the LCEs associated with a periodic orbit is always zero.

---

[5] A trivial corollary of the result above is that the LCEs associated with an attracting fixed point are all negative.

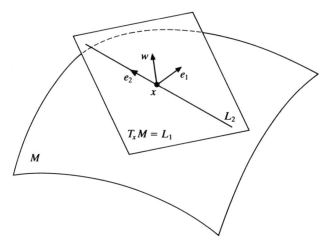

Fig. 6.2. Tangent space of the Lyapunov exponents χ_1 and χ_2 for a two-dimensional flow. For any tangent vector w that does not lie on the line $L_2, \chi(w, x) = \chi_1 \equiv \chi(e_1, x)$.

In principle, by appropriately choosing the vector w, we can also evaluate all the other LCEs.

In the general case in which orbits are not periodic, we cannot resort to the concepts of eigenvalues or eigenvectors since now $D\phi_t(x) : T_x M \to T_{\phi_t(x)} M \neq T_x M$. Nevertheless, it can be shown that the basic features of LCEs as described above are retained. In particular, when w varies in $T_x M$, it is still possible to find n numbers[6] obeying definition (6.4). It is also possible to define a sequence of linear subspaces $T_x M = L_1 \supset L_2 \supset \ldots \supset L_{n+1} = 0$, such that

$$w \in L_i(x) \backslash L(x)_{i+1} \iff \chi(w, x) = \chi_i(x),$$

where, if e_i is a basis of $T_x M$, $\chi_i(x) \equiv \chi(x, e_i)$; and $\in A \backslash B$ means 'belongs to A but not to B'. The ordered numbers $\chi_1(x) \geq \chi_2(x) \geq \ldots \geq \chi_n(x)$ constitute the *spectrum of LCEs* at x. It remains true that, if we choose at random a vector w of $T_x M$, $\chi(w, x) = \chi_1(x)$. (See Fig. 6.2.)

First-order LCEs tell us how lengths (i.e., distances between points) vary under the dynamic process. This notion can be generalized in order to consider the stretching and shrinking of *regions* of phase space. Thus,

[6] For simplicity's sake, we have assumed here that $\dim L_i(x) - \dim L_{i+1}(x) = 1$, i.e., there are no multiplicities. If there are, the formulae above are modified accordingly.

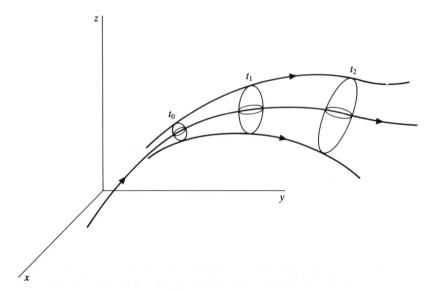

Fig. 6.3. Divergence of orbits originating from a small sphere of initial conditions.

we can define *LCEs of order p* as follows:

$$\chi(x, E^p) = \lim_{t \to \infty} \frac{1}{t} \ln Vol^p(D\phi_t(x)(U)), \qquad (6.5)$$

where $U \subset E^p \subset T_x M$ is a parallelepiped generated by p linearly independent vectors of E^p, and $Vol^p(A)$ is the p-dimensional volume of A induced by a scalar product $\langle,\rangle$ defined on A. If we put $Vol^p(U) = 1$, we can also write

$$|\det(D\phi_t(x)(U))| = Vol^p(D\phi_t(x)(U))$$

which is intuitively appealing, since a map can be said to be contracting or expanding according to whether the absolute value of the determinant is smaller or greater than one (and consequently its logarithm is smaller or greater then zero).

An interesting relation has been proven to exist between LCEs of order $p > 1$ and those of order one, namely: each LCE of order p is the sum of certain LCEs of order one, associated with certain linearly independent vectors $w_i \in E^p, i = 1, 2, \ldots, p$. In particular, if for a given x, we choose E^p at random, we have

$$\chi(x, E^p) = \chi_1(x) + \ldots + \chi_p(x), \qquad (6.6)$$

where the numbers χ_i are the p *largest* LCEs of order one.

6.2 LYAPUNOV CHARACTERISTIC EXPONENTS AND CHAOS

A positive LCE of order one implies that two nearby trajectories exponentially diverge (at least locally). One or more positive LCEs of order p indicate that, under the action of a flow (or a map), a small volume of initial conditions of dimension p gets mapped into a larger volume at a later time. This, in turn, implies that at least one of the LCEs of order one is positive (i.e., there is at least one direction along which we have SDIC), and that it dominates those which are negative. Equation (6.6) can also be very useful in numerical computations, especially when the dimension of the problem is small. Negative LCEs indicate contraction along certain directions, and zero LCEs indicate that along the relevant directions there is neither expansion nor contraction. In applications we are interested in bounded systems, i.e., those in which the values of variables remain finite. For these systems the (average) divergence of orbits can only be *locally* exponential, whereas *globally* there must be a process of 'folding' which merges widely separated points and keeps the motion bounded. It is precisely this combination of stretching (measured by positive LCEs) and folding that produces chaos.

We have seen that the calculation of LCEs may be thought of as an average over *time* (or, if the system is a discrete-time one, over *iterates* of the mapping). The limit of large t in (6.4) is necessary to describe long-term behaviour of the system and, if the motion takes place on an invariant set, to find a quantity independent of initial conditions (except perhaps a negligibly small set). If a group of solutions of a dynamical system is asymptotically attracted to a subset A of the phase space carrying an invariant measure μ which is ergodic, then it can be shown that LCEs are μ-almost everywhere constant, i.e., they are independent of the initial conditions. Moreover, in this case the 'time average' (6.6) can be replaced by a 'space average' over the phase space. In other words, we compute the local stretching, determined by the mapping $D\phi_t(x)$ (which depends on x), weighted by the probability of encountering that amount of stretching, that is the probability of a trajectory visiting a particular location x. Thus, the (first-order) LCE can be calculated as

$$\chi(w) = \int_M \mu(x) \ln \frac{\|D\phi_t(x)w\|}{\|w\|} d\mu,$$

where the integral is taken over the attractor of the flow, and, to simplify the issue, we have assumed that μ is absolutely continuous.[7] Analogously we can calculate a LCE of order p.

In discrete-time dynamical systems, the combination of local stretching and global folding which generates chaos can take place in one dimension only for non-invertible maps such as that investigated in Chapter 11. For invertible maps, the minimum dimension for chaos to occur is two. In continuous-time dynamical systems, chaos can take place only in three (or higher) dimensions.

The sign patterns of LCEs can be used to classify the different types of attractors:

(i)　　fixed point: $(-,-,\cdots,-)$

(ii)　　limit cycle: $(0,-,\cdots,-)$

(iii)　　T^n torus: $(\underbrace{0,\ldots,0}_{n\ \text{times}},-,\ldots,-)$

(iv)　　strange, or chaotic attractor: $(+,\cdots,+,0,-,\cdots,-)$

Notice that for attractors of continuous-time dynamical systems other than a fixed point, one of the LCEs must always be zero. This corresponds to the motion along the direction of the flow.

Remark 6.1.　　There is an interesting relationship between the LCEs of a flow, χ_i, and those of a first-return map derived from it, $\bar{\chi}_i$, namely:

$$\chi_i = \frac{\bar{\chi}_i}{\langle \tau \rangle_\mu},$$

where $\langle \tau \rangle_\mu$ indicates the average time between two intersections, computed with respect to the measure μ, i.e., the density of intersections on the surface of section.[8] Of course, the flow will have one characteristic exponent more than its Poincaré map, but this corresponds to the motion across the surface of section and, as we have already mentioned, it is always equal to zero.

Remark 6.2.　　As indicated in Chapter 2, if μ is an absolutely continuous (with respect to the Lebesgue measure) invariant measure carried by the attractor of a flow or a map, then the entropy $h(\mu)$ of the system is equal to the sum of the positive LCEs evaluated on the attractor. In fact, this equality also holds if μ is absolutely continuous along the unstable directions, corresponding to positive LCEs (see, section 2.8 above).

---

[7] See, however, the discussion of SRB measures in section 2.8 above.
[8] Cf. Ruelle (1989, p. 52); see also remark 4.3 above.

6.3 NUMERICAL COMPUTATION OF LCEs

The attempt to numerically compute LCEs by simply integrating the original differential equation and the variational equation, and then calculating LCEs according to the definition discussed above, runs up against two main problems.

Firstly, when at least one LCE is greater than zero and if we choose the tangent vectors w_i at random, straightforward application of the formula of equations (6.4) or (6.5) leads, after a sufficiently long period of time, to a computer overflow, as the norms of the vector w_i or the volume Vol^p increase exponentially. Secondly, the angles between the *directions* followed by the tangent vectors tend to zero and soon become indistinguishable, as those directions rapidly collapse to the dominant one, associated with the largest LCE.

These difficulties can be overcome, however, in view of the linear nature of the variational equation, by means of a known technique, called Gram–Schmidt re-orthonormalization procedure (GSR). Given a set of linearly independent vectors, GSR transforms it into a new set of conveniently normalized vectors which are orthonormal and preserve the orientation of the subspaces of the original set. In this way, both the divergence problem and the orientation collapse problem are avoided, and the calculation of the LCEs is not affected in any essential way.

The calculations performed by our software program can be summarized as follows:

(i)   When defining a system the user must provide a Jacobian matrix. This, of course, is obtained by taking partial derivatives of the vector field of the system. The elements of this matrix will be, in general, functions of the variables. As is well known, the Jacobian matrix evaluated *at equilibrium*, fully determines the qualitative behaviour of the system near it, provided that the equilibrium is hyperbolic, i.e., the matrix has no eigenvalue with zero real part, or, in the case of maps, no eigenvalue of the relevant matrix has modulus equal to one.

(ii)   The program introduces n new auxiliary variables for each of the original variables. A new $(n \times n)$ matrix Q is then formed by arranging on each row the auxiliary variables associated with the same original variable.

(iii) A product matrix JQ is defined, whose columns can be thought of as dynamic linear systems, each of which is associated with a certain displacement from the reference point x, on the tangent space $T_x M$. When dealing with ordinary differential equations, the original system as well as the n auxiliary linear systems are integrated by means of the standard routines employed by the program.

(iv)   The 'divergence problem' is overcome by replacing the vectors of co-ordinates produced at each step by the integration (or iteration) procedure, by corresponding normalized vectors. Due account must be taken of the size of the normalization in order to evaluate the LCEs correctly.

(v)   The 'orientation problem' is dealt with by applying a standard GSR routine and recalling that, given the linear nature of the map $D\phi_t(x)$, the ratio $Vol^p(D\phi_t(x)(U))/Vol^p(U)$ is independent of U.

6.4 COMPUTATION OF LCEs FROM EXPERIMENTAL DATA

The estimation of LCEs from experimental data can be performed by essentially the same method as for data derived from a known model.[9] There are some important differences, however. Consider the simpler case of the largest Lyapunov exponent. One begins by choosing a 'fiduciary trajectory' (f.t.) and a nearby 'test trajectory' (t.t.). One then wishes to evaluate the local rate of divergence between the two trajectories, averaged over the attractor. Now, when the equations of motion (i.e., the original model) are available, t.t.s are naturally supplied by (periodically renormalized) vectors in the tangent space at each point of the f.t.

This approach is not directly applicable to experimental data as here no equations of motion are available. Instead, two initial points are chosen as near one another as possible, but whose temporal separation in the time series is no less than one mean orbital period. So long as the spatial separation is small, the vector between these two initial points can be taken as an approximation to a vector in the tangent space. In the process of renormalization, we have to pick up another point near the f.t., and possibly with the same orientation as the original vector.

However, working with experimental data, the number of points available is limited and things can go wrong, either because the minimum

---

[9] A more detailed discussion of the general problem of detecting deterministic chaos in real time series will be developed below in Chapter 10.

distance available is too large or because of distortion in the orientation. Clearly, for any given kind of attractor, much depends on the quantity and quality (e.g., degree of noisiness) of the data available, as well as the skill of the practitioners.

Our software program includes two subroutines to reconstruct an attractor from a uni- or multivariate time series, and to calculate the associated *dominant* LCE. The time series itself may be an external input, or may be the output of an integration of a dynamical system, performed by the program itself (in the latter case, we shall speak of 'pseudo-experimental data').

The 'reconstruction' procedure follows directly the steps indicated in the theoretical discussion above and does not pose any particular computational problems. The difficult choices are all the user's, who must provide a guess on the dimension of the 'pseudo-phase space' and choose the time delay.

The subroutine for calculation of the dominant LCE from a time series is essentially the same as that produced by Wolf *et al.* (1985, pp. 312–13), to which we refer the user for a more detailed discussion of the problems outlined above.

A simple, though possibly risky, method to evaluate the largest LCE from experimental data is available, namely the already mentioned 1D map approach. Suppose we have already successfully reconstructed the low-dimensional attractor of an unknown dynamical system, and that the attractor 'lives' in, say, a three-dimensional space. Let us next construct a two-dimensional Poincaré section, transverse to the attractor, so that now, instead of a long curve in three dimensions, we have a set of points S lying on a two-dimensional surface Σ, which for simplicity's sake we take to be a plane. If S approximates a one-dimensional ordering of points, we are encouraged to look for a one-dimensional map. For this purpose, we calculate the distance of each of the points in S from one or the other end of Σ. Then we plot the distance of each point against that of its successor in time. In a number of cases, the plot approximates a well-defined one-dimensional map of a known type, for example a unimodal map. If so, we can fit the computed co-ordinates to an appropriate function (or interpolate them), and then numerically evaluate the single LCE.

Although care must be taken in applying this technique (especially when dealing with with experimental data), it is known that this procedure sometimes produces surprisingly accurate results. In particular, it is very useful in estimating the dominant LCE, which can be easily calculated from the 1D map.[10] A successful application of this method can be seen in Chapter 11 below.

---

[10] On this point, cf. Wolf and Swift (1984), Simoyi *et al.* (1982), Thompson and Stewart (1986, pp. 239–45); Tomita (1986, pp. 224–34).

7

Dimensions

7.1 FRACTAL DIMENSION

In the previous chapter we have given a basic theoretical background and described a few practical techniques for the calculation of Lyapunov characteristic exponents. The concept of 'fractal dimension' provides a second important tool, by means of which we can attempt to give a quantitative characterization of strange or chaotic attractors. It may also be used, at least in principle, to distinguish between deterministic chaos and random motion.

An intuitive and unsophisticated notion of dimension in the context of dynamical system theory is given by the Euclidean dimension of phase space, or more simply by the number of phase variables. However, for a deeper understanding of the behaviour of dynamical systems, subtler and more powerful concepts are needed, in particular that of 'fractal dimension', which is so often associated with chaotic dynamics. We shall start with a fundamental idea developed at the beginning of this century by Hausdorff and Besicovitch.

Let S be a set of points in a space of Euclidean dimension p. (Think, for example, of the intersection points generated by a Poincaré first-return map and lying on a p-dimensional surface of section.) We now consider certain hypercubes of side ϵ (or, equivalently, certain hyperspheres of radius ϵ), and calculate the minimum number of such cells, $N(\epsilon)$, necessary to 'cover' S.

Then, the fractal dimension D of the set S will be given by the following limit (assuming it exists):

$$D \equiv \lim_{\epsilon \to 0} \frac{\log(N(\epsilon))}{\log(1/\epsilon)}. \tag{7.1}$$

To be precise, the quantity defined in (7.1) is called the (Kolmogorov) *capacity dimension*. However, the concepts of capacity (D) and Hausdorff

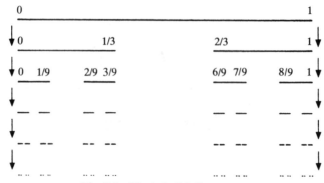

Fig. 7.1. The 'triadic' Cantor set.

(D_H) dimension are closely related.[1] It can be proved that $D_H \leq D$, strict equality holding in some special cases, and the two are sometimes confused in the literature. For our present purposes, however, the simpler and more intuitive formula (7.1) will suffice and, whenever the term 'dimension' is used without qualifications, we shall be referring to that formula. Notice that both the capacity and the Hausdorff dimensions are metric, not probabilistic measures, as they are independent of the relative densities of points in different regions of a set.

It is easily seen that, for the most familiar geometrical objects, (7.1) provides perfectly intuitive results. For example, if S consists of just one point, $N(\epsilon) = 1$ and $D = 0$; if it is a segment of unit length, $N(\epsilon) = 1/\epsilon$, and $D = 1$; finally, if S is a plane of unit area, $N(\epsilon), = 1/\epsilon^2$ and $D = 2$, etc. That is to say, for 'regular' geometrical objects, dimension D does not differ from the usual Euclidean dimension, and, in particular, D is an integer. This is not always true, however, as the well-known example of the 'triadic' Cantor set will show. The basic idea can be understood by looking at Fig. 7.1.

Consider again a segment of unit length; divide it into three equal subsegments and remove the intermediate one: you will be left with two segments of length 1/3. Divide each of them into three segments of length 1/9 and remove the (two) intermediate ones. Now iterate n times and then pose the question: what is the dimension of the resulting geometrical object, C? For n finite, C is a collection of segments and its dimension is clearly one. But in the limit, for $n \rightarrow \infty$, C is an infinite

---

[1] A precise definition of the Hausdorff dimension can be found in Guckenheimer and Holmes (1983, p. 285); see also Eckmann and Ruelle (1985, p. 620).

collection of points and the answer may not be obvious. To verify this, let us try to evaluate the limit (7.1), proceeding step by step. Consider first a (one-dimensional) cube of side ϵ. Clearly we shall have $N(\epsilon) = 1$ for $\epsilon = 1$, $N(\epsilon) = 2$ for $\epsilon = 1/3$, and, generalizing, $N(\epsilon) = 2^n$ for $\epsilon = (1/3)^n$. Taking the limit for $n \to \infty$ (or equivalently taking the limit for $\epsilon \to 0$), we can write

$$D = \lim_{\substack{n \to \infty \\ (\epsilon \to 0)}} \frac{\log 2^n}{\log 3^n} \approx 0.63. \tag{7.2}$$

We have thus characterized quantitatively a geometric set that is more complex than the usual Euclidean objects. The dimension of C is a non-integer. We might say that C is an object 'greater' than a point but 'smaller' than a segment.

Notice, however, that the length of C, L, is zero. Indeed, we have $L = 1 - (\frac{1}{3} + \frac{2}{9} + \ldots + \frac{n}{3^n})$ and $\lim_{n \to \infty} L = 0$. From our discussion of section 2.8 above, we know that this implies that the Lebesgue measure of C is zero. Thus, the capacity dimension allows us to measure (and compare) objects whose Lebesgue measure is zero.

The Kolmogorov capacity and the Hausdorff dimension are prototypical members of a family of dimension-like quantities which can be grouped under the label 'fractal dimensions' and can be regarded as measures suitable for fractal[2] objects, i.e., objects characterized by a non-integer dimension. When applied to geometrical analysis of dynamical systems, the concept of fractal dimension can be conceived of as a measure of the way orbits fill the phase space under the action of a flow (or a map). A non-integer fractal dimension indicates that orbits of a system tend to fill up less than an integer subspace of the phase space. The relevance of this fact for analysis of chaotic dynamics arising from iterations of maps can be appreciated by considering that non-invertible, one-dimensional maps, for example (like those discussed in Chapters 9 and 11), under certain conditions on the controlling parameter, are characterized by invariant sets (which may or may not be attracting) which are Cantor sets. Two-dimensional maps (like that investigated in Chapter 12) may have invariant sets of the 'horseshoe type' (cf. Smale, 1967 and Chapter 8), which, geometrically, are the products of Cantor sets.

The association between non-integer fractal dimension and chaos for

---

[2] The term 'fractal' was coined by Benoit Mandelbrot and it refers to geometrical objects characterized by 'self-similarity', i.e., objects having the same structure on all scales. Each part of a fractal object can thus be viewed as a reduced scale of the whole. Intuitively, a snowflake can be taken as a natural fractal. The Cantor set, discussed above, gives a mathematically more precise example.

dissipative, continuous dynamical systems can be understood by considering that their complex dynamics are the result of the infinitely repeated stretching and folding of a bundle of orbits under the action of the flow. To visualize this idea in a non-rigorous, but intuitively appealing manner, let us consider a flow occurring in a bounded subset of a three-dimensional phasespace constructed in the following way[3] (see Fig. 7.2).

Orbits of the flow move largely in one direction (indicated by the big arrow); they tend to spread out in a second direction (intermediate arrow) and they are pushed together in a third one (small arrows). As a result of this compression, a three-dimensional volume of initial conditions will be squeezed onto a (quasi) two-dimensional 'sheet'. But, although nearby orbits diverge exponentially, the motion of the system, by construction, is bounded. Therefore the 'sheet' on which the flow takes place, though continuously expanded along its width, must be folded over on itself and, for the same reason, the two ends AB, $A'B'$ of the 'sheet' must be twisted and smoothly joined together. However, since $A'B'$ has two distinct 'sheets' and it joins AB which has only one, in order for the joining to be smooth (which is a necessary condition for the system to be representable by a continuous invertible flow), the 'sheet' on which the flow occurs must in fact be a 'book' of thickness between 0 and 1, made of infinitely many 'leaves', each of zero thickness. In fact, if we position a two-dimensional Poincaré surface of section across the orbits, the resulting set of intersection points will have a fractal structure similar to that generated by the 'horseshoe map'. As we have mentioned in Chapter 4 in the discussion of Poincaré maps, when the compression (small arrows) is very strong, the resulting Poincaré map may look one-dimensional.[4]

A mechanism broadly corresponding to the one just described seems also to be at the root of the strange behaviour of the systems discussed below in Chapters 11 and 13. The geometrical aspect of the attractors of these systems, as well as the relevant numerical calculations, indicate that their fractal dimensions are non-integer and between 2 and 3.

In fact, it can be easily shown that chaotic attractors of three-dimensional, continuous-time dissipative systems must have a fractal dimension $2 < D < 3$. To see this, let us consider that the volume of any initial three-dimensional set of initial conditions asymptotically shrinks under the action of the flow. This means that the dimension D of any attractor

---

[3] Cf. Shaw (1981, pp. 94–5); also Lichtenberg and Lieberman (1983, pp. 57–9).
[4] See, on this point, Cvitanovic (1984, p.18).

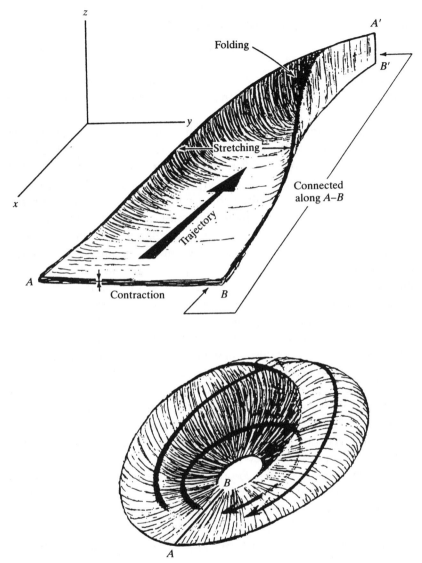

Fig. 7.2. Stretching and folding in a chaotic attractor.

of the system must be smaller than 3. On the other hand, for the attractor to be chaotic (in the sense that it possesses SDIC), its dimension must be greater than 2; no attractor more complex than a limit cycle may occur in $\mathbb{R}^2$ (and nothing more complex than a quasiperiodic orbit may occur in T^2). Consequently, it must be the case that

$$2 < D < 3.$$

An example of a system generating a flow in a ten-dimensional phase space and characterized by a strange attractor with dimension D, $2 < D < 3$ is given in Chapter 11.[5]

A special comment is deserved by the chaotic sets occurring in one-dimensional maps of the logistic type, discussed in greater detail below in Chapter 9. From our discussion of Chapter 2 it can be gathered that, if those sets support an invariant ergodic measure which is absolutely continuous with respect to the Lebesgue measure (i.e., if chaos is observable), their fractal (capacity) dimension must be equal to one. Therefore, the sets in question are not attractors according to our definitions (2.6) and (2.7). In particular, they do not have any 'shrinking neighbourhood' and consequently they are not attracting sets. The same idea could also be illustrated by saying that one-dimensional discrete-time dynamical systems with a positive LCE (i.e., chaotic) cannot be dissipative.

The concept (and the measurement) of fractal dimensions are not only necessary to understand the finer geometrical structure of strange attractors, but they are also fundamental tools for providing quantitative analysis of such attractors. Since the latter are mathematical objects which represent the permanent regime of dissipative dynamical systems, qualitative and quantitative information concerning attractors is of paramount importance in the understanding of the overall nature of the flow generated by the system. A particularly valuable piece of information concerns the level of complexity of the flow, as indicated by the number of 'essential' co-ordinates of the system. In other words, although originally the number of the phase space variables may be high, the permanent dynamics often take place on certain subsets of the phase space with a much lower dimension. Thus, the dimension of the attractor of the system (e.g., as measured by the formula (7.1)) can be taken as

---

[5] A similar case of a fifty-dimensional system of o.d.e. with an attractor of dimension $2 < D < 3$, is discussed in Invernizzi and Medio (1991). For a more general argument, saying that all strange attractors must have non-integer fractal dimensions, see Helleman (1980, in Cvitanovic, 1984, p. 461). On the other hand, a non-integer fractal dimension does not imply chaotic dynamics in the sense of sensitive dependence on initial conditions. For example, the so-called Feigenbaum attractor (see Chapter 9 below) has a non-integer fractal dimension, but not SDIC, i.e., it is not chaotic. Grebogi *et al.* (1984) have provided other examples of dynamical systems for which non-integer fractal dimensions do not imply SDIC.

an index of complexity or, in the case of strange attractors, an index of strangeness.[6]

Remark 7.1. Notice that a complex geometrical structure (as indicated by a non-integer fractal dimension) may characterize not only the attractors of a system, but also their basic boundaries, i.e., the boundaries which separate the different zones of the phase space 'controlled' by different attractors. In applications, this may greatly complicate the problem of steering the system towards a course that is more desirable than others.

7.2 NUMERICAL COMPUTATIONS

As in the case of LCEs, analytical computation of the fractal dimension is restricted to a small number of cases. For the great majority of systems of practical importance, we must rely on numerical computation.

Unfortunately, fractal dimension as defined by (7.1), in practice (i.e., whenever $D > 2$) cannot be computed easily, and convergence of the limit may not be guaranteed. A different approach has been suggested by Grassberger and Procaccia (1983). The basic idea is that of replacing the so-called 'box counting' algorithm, necessary to compute $N(\epsilon)$ in the formula (7.1), with the measurement of distances between points representing positions of the system along an orbit on the attractor. As the Grassberger and Procaccia method is at the moment prevailing in applications, and as it is also the method adopted in the software program attached to this book, we shall describe its basic procedures in some detail.

Suppose we start from the usual system of differential equations

$$\dot{x} = f(x), \qquad x \in \mathbb{R}^n.$$

By numerical integration we obtain a trajectory consisting of N discrete points $\{x_i\}_{i=1}^N \equiv \{x(t+i\tau)\}_{i=1}^N$, where τ is an arbitrarily fixed time interval, and the initial points have been selected so as to discard the transients and (hopefully) to be able to study the motion of the system on its attractor.[7]

---

[6] We shall return to this point in Chapter 10, with regard to the analysis of experimental signals.

[7] When one is analysing an experimental signal, and the system which has generated it is unknown, one may create a pseudo-phase space, along the lines discussed in Chapter 10. What follows concerns both the calculations based on the output of a given model, and those based on experimental signals.

More rigorously, we define the *correlation function*

$$C(r) = \lim_{N \to \infty} \frac{1}{N^2} \sum_{\substack{i,j=1 \\ i \neq j}}^{N} \theta(r - |x_i - x_j|),$$

where $\theta(s)$ is the Heavyside function, i.e.,

$$\theta(s) = \begin{cases} 1 & \text{if } s \geq 0 \\ 0 & \text{if } s < 0. \end{cases}$$

Grassberger and Procaccia argue (and for many attractors numerical evidence has been found) that, for small r, $C(r)$ behaves as a power of r. If so, we can write

$$C(r) \propto r^{D_c},$$

where D_c will henceforth be called the 'correlation dimension'. Clearly this function is limited to values of r which are small relative to the size of the attractor, since, for sufficiently large r, $C(r)$ tends to 1, independently of r (effect of saturation: all points on the attractor are counted). On the other hand, for sufficiently small r, $C(r)$ tends to zero (no point is counted). In numerical estimates, in which the number of points is limited and precision is finite, for too small r the estimate statistics are poor, since the number of observations is small and errors become important. In practice, one tries to define a 'scaling region' with lower and upper bounds (in terms of r), over which the slope of $\log C(r)$ versus $\log r$ curve is approximately constant. Examples of this procedure can be seen in Chapters 11–13 (cf. Figs. 11.7; 12.9 and 12.15; 13.8).

In the computation of the correlation dimension, we have employed an algorithm due to Parker and Chua (1987 and 1989, pp. 179–91), which takes advantage of the floating point representation used in computers, and is much faster than the other algorithms most commonly used. Notice that with this method the values of r over which the computations will be performed are not chosen by the user but are determined endogenously. For details we refer the reader to the works cited.

It has been shown (Grassberger and Procaccia, 1985) that the following inequality holds:

$$D_c \leq D,$$

i.e., the correlation dimension is a lower bound for the capacity dimension.

Notice that D_c (unlike D and D_H) is a probabilistic, not a metric dimension, i.e., it accounts for the frequency with which an orbit visits the different parts of the attractor. If the covering of the attractor is uniform, the above inequality is replaced by an equality and the two measures of dimension coincide. For many known attractors, however, even though those measures differ, their estimates have been found very close to one another. For systems whose equations of motion are known (and with a reasonably low number of phase space variables), correlation dimension can be computed fairly easily and convergence is often very rapid.

For experimental systems, however, when the time series of only one or a few variables are available, the situation is more difficult. The method usually followed is related to the procedure discussed in Chapter 10, to which the reader is referred for details. In practice, one must first of all make a guess concerning the dimension of the (assumed) attractor of the (unknown) system which has supposedly generated the time series. Then, one 'embeds' the time series in that dimension and tries to reconstruct the attractor and estimate its correlation dimension accordingly. The procedure is iterated for progressively higher dimensions. In principle, insofar as the dimension of the assumed attractor, D_c^a, is smaller than that of the 'true' attractor $\bar{D}_c$, the D_c^a calculated will continue to increase. If the system is deterministic, however, it will converge to $\bar{D}_c$. In the opposite case, in which the motion of the system is purely random and the analysed signal is white noise, $\bar{D}_c = \infty$ and therefore the calculated dimension will increase indefinitely.

This fact, incidentally, suggests the possibility of using correlation dimension estimates in order to distinguish between deterministic and stochastic systems. In practice, however, one always deals with time series which, although conceivably generated by deterministic systems, are 'contaminated' by noise and errors of various kinds, so that a clear-cut answer to the problem at hand may be very difficult or even impossible. More on this point can be found in Chapter 10.

7.3 OTHER MEASURES OF FRACTAL DIMENSION

As already mentioned, there exist other criteria for measuring fractal dimension, and more may be suggested in the future. We shall briefly list here some of the best known ones.

The *information dimension* can be considered the probabilistic counterpart of the capacity dimension. As we did in our discussion of capacity dimension, let us cover the set of points whose dimension we want to measure with a set of $N(\epsilon)$ cells of size ϵ. Next, let us compute the number of points N_i $(i = 1, \ldots, n)$ in each of the $N(\epsilon)$ cells, as well as the probability p_i of finding a point in each of the cells. Then, we can write

$$I(\epsilon) = - \sum_i^{N(\epsilon)} p_i \log p_i,$$

which is the so-called *information entropy*, studied by Shannon (1948). This is a measure of the uncertainty about the system. In Chapter 2 we discussed a generalization of this concept and the resulting notion of (Kolmogorov–Sinai) entropy, as a measure of the 'disorder' of the system.

For small ϵ, it is found that $I(\epsilon)$ behaves as

$$I(\epsilon) \approx D_I \log(1/\epsilon),$$

so that, for small ϵ, we can define the information dimension thus:

$$D_I = \lim_{\epsilon \to 0} \frac{I(\epsilon)}{\log(1/\epsilon)}.$$

In general, it can be shown that

$$D_I \leq D.$$

Clearly, if the covering of the attractor were uniform, we should have $p_i = 1/N \ \forall \ i$, N being the total number of points. Consequently, in this case, we should have

$$I(\epsilon) = \log N(\epsilon),$$

and

$$D_I = D.$$

It has also been shown that, in general, we have

$$D_c \leq D_I \leq D,$$

so that the combined computation of capacity and correlation dimension could provide an excellent estimate of the information content (degree of predictability) of a strange attractor (cf. Grassberger and Procaccia, 1983).

Yet another probabilistic measure of fractal dimension is given by the *pointwise dimension*, defined as

$$D_p = \lim_{r \to 0} \frac{P(r, x_i)}{\log(r)},$$

$P_i(r, x_i)$ being the probability of finding a point (belonging to a set S) in a sphere of radius r, centred at a point of the phase space $x_i \in S$.

In general, however, P will depend on x_i and an averaged pointwise dimension may have to be used (e.g., calculating it over a large number of randomly chosen points).

Finally, special consideration must be given to the so-called *Lyapunov dimension*, whose simple computation is based on the knowledge of LCEs, which we have discussed in Chapter 6. Suppose we have computed the n LCEs $(\lambda_1, \lambda_2, \ldots, \lambda_n)$ evaluated on the attractor of a certain dynamical system, and we have listed them in decreasing order, thus:

$$\lambda_1 > \lambda_2 > \ldots > \lambda_n.$$

Suppose also that s is the largest number for which

$$\sum_{i=1}^{s} \lambda_i > 0,$$

and that

$$\lambda_1 + \lambda_2 + \ldots + \lambda_{s+1} < 0.$$

Then it must be that

$$\lambda_{s+1} < 0, \qquad \frac{\lambda_1 + \lambda_2 + \ldots + \lambda_s}{|\lambda_{s+1}|} < 1.$$

It can be proved (cf. Temam, 1988, pp. 287–8) that

$$D_H \leq s + \frac{\sum_{i=1}^{s} \lambda_i}{|\lambda_{s+1}|} \equiv D_L.$$

Although it has also been conjectured by Kaplan and Yorke (1979) that, generically:

$$D_H = D_L,$$

it is not clear yet under what concept of generality their conjecture may be proved. In most known applications, however, D_L gives a surprisingly good approximation to D (and therefore to D_H) and it may provide all the information we need. When the equations of motion are known and all the LCEs can be computed, derivation of D_L is elementary. (For an application to an economic model, see Chapter 11.)

However, in the analysis of experimental signals, the usefulness of the formula above is more dubious since, in this case, it is not easy to compute the LCE λ_{s+1} associated with contraction.[8]

---

[8] For a relation between fractal dimension and power spectrum, see Schuster (1989, pp. 56–9).

8

Symbolic dynamics

8.1 THE SPACE OF SYMBOL SEQUENCES. THE SHIFT MAP

In this chapter, we shall discuss the possibility of defining sufficient conditions for certain dynamical systems to exhibit chaotic behaviour. The method in question is one of the very few tools available to establish the presence of chaos analytically as opposed to numerically. To be more precise, the application of this method enables us to prove that certain maps possess invariant sets with very complicated structures.

Even though this result does not imply the existence of chaotic *attractors*, the method in question is interesting for two reasons. Firstly, it gives us an illuminating picture of certain basic mechanisms which are at the root of complex behaviour. Secondly, proving the existence of chaotic invariant sets may sometimes be the first step towards proving that a system possesses a chaotic attracting set or a chaotic attractor.[1]

Broadly speaking, the strategy will be first to formulate the problem in terms of a map, then to construct a second auxiliary map for which the dynamics can be proved to be chaotic (in a sense to be defined) and, finally, to establish the conditions for a certain topological equivalence between the original and the auxiliary map. If the problem to be investigated is originally formulated in terms of differential equations, a preliminary step will be necessary, in order to conveniently replace the flow with a map.

The method is neither simple nor easy to present in a nontechnical manner. Keeping with the tenor of this book, we shall discuss it concisely (omitting all proofs), but with sufficient precision of language, and we shall refer the reader to the specialized literature for a more detailed

---

[1] Cf. on this point, Wiggins (1990, p. 612).

discussion.[2] The first step in the reasoning is to briefly describe a technique for characterizing the orbit structure of a dynamical system via infinite sequences of 'symbols', which is known as 'symbolic dynamics'.[3] This concept is rather abstract and at first it may be difficult to grasp it intuitively; but this is not required in the present context. Only the logic of the argument needs to be understood here.

Let $S = \{1, 2, 3, \ldots, N\}$ denote a collection of symbols. In a physical interpretation, the elements of S could be anything, for example letters of an alphabet, or isolated readings of some measuring device from the observation of a given dynamical system. Here we shall use positive integers and assume that S is finite so that $N \geq 2$ is a fixed positive integer.

Next, from the elements of S we want to construct the space of all symbol sequences, which we shall denote by Σ and we define as

$$\Sigma^N \equiv \ldots \times S \times S \times S \times S \times \ldots \equiv \prod_{i=-\infty}^{+\infty} S_i,$$

where $S^i = S \ \forall \ i$.

A point s in Σ^N is therefore represented as a 'bi-infinity-tuple' of elements of S, i.e.,

$$s \in \Sigma^N \Rightarrow s = \{\ldots, s_{-n}, \ldots, s_{-1}, s_0, s_1, \ldots, s_n, \ldots\},$$

where $s_i \in S \ \forall \ i$.

As a simple example, let us put $N = 2$ so that $S = \{1, 2\}$, and $s \in \Sigma^2$ is a bi-infinite sequence of 1s and 2s, such as

$$s = \{\ldots, 1, 1, 2, 1, 2, 2, 1, 2, 1, \ldots\}.$$

It can be proved that:

(i)   It is possible to define a metric on S and consequently on Σ^N. Since the elements of S are positive integers, we can define the distance between two elements of S as the absolute value of their difference, i.e.,

$$d(a, b) \equiv |a - b| \qquad \forall \ a, b \in S.$$

---

[2] In what follows, we shall draw upon Wiggins (1988), which contains the most thorough and detailed discussion of which we are aware at the moment.

[3] The origin of the idea goes as far back as the end of the nineteenth century, when it was applied by Hadamard (1898). After a long period of neglect, symbolic dynamics has experienced, from the mid-1960s onwards, a tremendous comeback with numerous applications. Besides the book by Wiggins, already cited, cf. Moser (1973).

Similarly for Σ^N the distance between s and $\bar{s}$ will be defined as

$$d(s, \bar{s}) \equiv \sum_{-\infty}^{+\infty} \left(\frac{1}{2^{|i|}} \right) \frac{|s_i - \bar{s}_i|}{1 + |s_i - \bar{s}_i|}. \tag{8.1}$$

(ii)   The space Σ^N endowed with metric (8.1) is compact, totally disconnected and perfect.[4] Notice that these properties are topologically invariant, i.e., they are invariant under homeomorphism. Also notice that they are taken as the defining properties of a Cantor set, of which the 'triadic' kind has been discussed in Chapter 7.

We shall now define a map $\sigma : \Sigma^N \to \Sigma^N$, called a *shift map*, as follows:

For $s = \{\ldots s_{-n} \ldots s_{-1}.s_0 s_1 \ldots s_n \ldots\} \in \Sigma^N$,

$$\sigma(s) \equiv \{\ldots s_{-n} \ldots s_{-1} s_0 . s_1 \ldots s_n \ldots\},$$

(notice the displacement of the 'decimal point'!), or more compactly:

$$[\sigma(s)]_i \equiv s_{i+1}.$$

The map σ, often referred to as a *full shift on N symbols*, shifts all entries in a sequence one place to the left. Now the following proposition can be proved:[5]

Proposition 8.1.   *The shift map σ acting on Σ^N is continuous and has:*

(*i*)   *a countable infinity of periodic orbits consisting of orbits of all periods; these periodic orbits are all of a saddle type;*

(*ii*)   *an uncountable infinity of bounded nonperiodic orbits;*

(*iii*)   *a dense orbit.*

Moreover, for any point $p \in \Sigma^N$, no matter how small a neighbourhood of p we consider, there is at least one point in this neighbourhood such that, after a finite number of iterations, the distance between p and this point (measured as indicated by (8.1)) is no less than some fixed value.[6] Therefore, we can write the additional proposition:

Proposition 8.1′.   *The shift map on Σ^N possesses* sensitive dependence on initial conditions, *hence it is chaotic.*

In some applications, one may wish to restrict the domain of σ so as to exclude certain possible symbol sequences. This is done by making

---

[4] A set is said to be *compact* if it is closed and bounded; *totally disconnected* if it contains no intervals; *perfect* if it is closed and every point in the set is a limit point of other points in the set.

[5] For a proof, see Wiggins (1988, p. 101).

[6] Cf. Wiggins (1988, p. 97).

use of a matrix A, called a 'transition matrix', whose elements are 0s or 1s, and which is defined by:

$(A)_{i,j} = 1$, if the ordered pair of symbols i, j is allowed to appear in the symbol sequence,

$(A)_{i,j} = 0$, if the contrary is true.

A simple example will clarify the point. Suppose once more that only two symbols are considered so that $N = 2$, $S = \{1, 2\}$. Suppose also that, for whatever reasons, we want to exclude all sequences $\{2, 1\}$ and $\{1, 2\}$ from the possible sequences of 1s and 2s allowing all the others. Then we shall have

$$A = \begin{pmatrix} 1 & 0 \\ 0 & 1 \end{pmatrix},$$

and the space

$$\Sigma_A^2 = \{s = \{\ldots s_{-n} \ldots s_{-1} s_0 s_1 \ldots s_n \ldots\} \in \Sigma^2 \mid (A)_{s_i, s_{i+1}} = 1 \; \forall \; i\}$$

consists of two points, i.e., the sequences $\{\ldots 111 \ldots\}$ and $\{\ldots 222 \ldots\}$.

The following propositions can be proved:

Proposition 8.2.   *Suppose A is irreducible; then Σ_A^N is compact, totally disconnected and perfect, i.e., it is a Cantor set.*

Proposition 8.3.   *Suppose A is irreducible. Then the shift map σ on Σ_A^N has the same properties as those described in Propositions 8.1 and 8.1'.*

Equipped with these ideas and propositions, let us now consider a continuous and sufficiently differentiable map

$$f : D \to \mathbb{R}^n,$$

where $D \subset \mathbb{R}^n$ is closed and bounded. Suppose that:

C.1 f possesses a non-empty invariant set Λ so that $f(\Lambda) = \Lambda$.

C.2   There exists a one-to-one correspondence between each point of Λ and an element of Σ_A^N (a bi-infinite sequence of symbols), such that we can define a homeomorphism $\phi : \Lambda \to \Sigma_A^N$ and a commutative diagram ('commutative' here means that $\phi \circ f = \sigma \circ \phi$)

$$
\begin{array}{ccc}
\Lambda & \xrightarrow{\;\; f \;\;} & \Lambda \\
\phi \downarrow & & \downarrow \phi \\
\Sigma_A^N & \xrightarrow{\;\; \sigma \;\;} & \Sigma_A^N
\end{array}
$$

Whenever the conditions C.1 and C.2 are fulfilled, we can conclude that the dynamics of f on Λ are essentially the same as those of

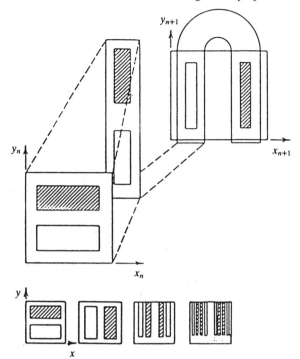

Fig. 8.1. Stretching and folding in the horseshoe map.

σ on Σ_A^N, and its orbit structure has the properties described by Propositions 8.1 and 8.1′. Hence we can conclude that Λ is *chaotic* in the precise sense that the dynamics of f on it possess sensitive dependence on initial conditions.

The difficult notion here is to imagine that a point in the $\mathbb{R}^n$ space may correspond to a bi-infinite sequence of symbols such as S. To understand this, one should consider that, to each initial state $x \in D$ there will correspond one infinite sequence $\{f(x), f^2(x), f^3(x), \ldots\}$ and, if f is invertible, a second infinite sequence $\{f^{-1}(x), f^{-2}(x), f^{-3}(x), \ldots\}$.

8.2 THE SMALE HORSESHOE AND THE GEOMETRY OF CHAOS

General considerations apart, there remains the problem of verifying the conditions on the map f, under which the hypotheses C.1 and C.2 above hold. Rather than discussing this problem in general, we shall give a brief presentation of its prototypical example, known as the 'Smale horseshoe'

(Smale, 1963), and then shall mention how the relevant results may be generalized.

Smale's is a simple map transforming a set of points in the plane, whose action can be described as follows.

Consider the map

$$f : D \to \mathbb{R}^2, \qquad D = \{(x, y) \in \mathbb{R}^2 \mid 0 \leq x \leq 1, 0 \leq y \leq 1\},$$

which acts on a unit length square D, contracting the x direction by a factor $1/a > 2$, expanding the y direction by a factor $1/b > 2$, twisting D around and laying it back on itself – as indicated in Fig. 8.1. Analogously, we can define the inverse action, i.e., $f^{-1}(D)$, as indicated in Fig. 8.2.

It will have been noticed that, under iteration of f (or f^{-1}), only certain subsets of D are mapped back onto D itself, whereas points outside those subsets will fall outside D.

In order to analyse the *recurrent* behaviour of the map, we must iterate f (and f^{-1}) an indefinite number of times, and try to identify a set $\Lambda \subset D$) such that $f(\Lambda)$ and $f^{-1}(\Lambda)$ are identical with Λ itself. In other words, we must construct the set

$$\ldots \cap f^{-n}(D) \cap \ldots f^{-1}(D) \cap D \cap f(D) \cap \ldots f^{n}(D) \cap \ldots,$$

or

$$\lim_{k \to \infty} \bigcap_{n=-k}^{k} f^{n}(D).$$

It will be expedient to split the invariant set Λ into Λ^+ and Λ^-, so that $\Lambda^+ \equiv \cap_{n=0}^{+\infty} f^{n}(D)$, $\Lambda^- \equiv \cap_{n=0}^{+\infty} f^{-n}(D)$ and $\Lambda = \Lambda^+ \cap \Lambda^-$.

The question now is: is the set Λ non-empty and if so what are its topological characteristics? From the geometric definitions of the maps f and f^{-1}, and applying them recursively, it is possible to give a complete answer to these questions, which we shall only summarize here.[7]

We have seen that $f(D) \cap D$ consists of 2 vertical bands across D of width a; likewise $f^2(D) \cap D$ consists of 2^2 vertical bands of width a^2; $f^3(D) \cap D$ consists of 2^3 vertical bands of width a^3, and so on. Inductively, we can find that $\Lambda_k^+ = \cap_{n=0}^{k} f^{n}(D)$ consists of 2^k vertical strips each of width a^k. In the limit as $k \to \infty$, we shall obtain an infinite number of vertical lines, whose projection on the horizontal side of the square D consists of an uncountable collection of points. In fact, if we put $a = 1/3$, this projection will be identical to the 'triadic' Cantor set which we have discussed in Chapter 7. Analogously, we find that $\cap_{n=0}^{k} f^{-n}(D)$ consists

---

[7] A detailed discussion can be found in Wiggins (1988, pp. 79–86).

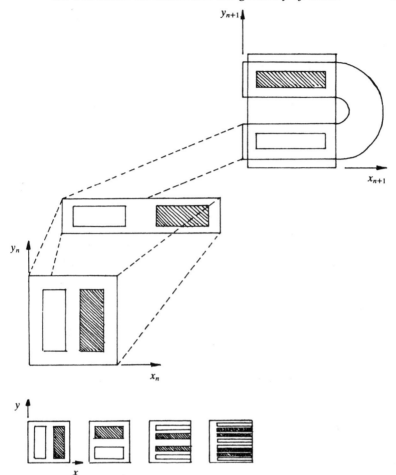

Fig. 8.2. Stretching and folding in the inverse horseshoe map.

of 2^k horizontal strips of height b^k and, letting $k \to \infty$, we conclude that Λ^- consists of an infinite number of horizontal lines, whose projection on the vertical side of D is a Cantor set. The set $\Lambda^+ \cap \Lambda^-$ is the product of two Cantor sets and is itself a Cantor set. An approximate (truncated) picture of Λ is given in Fig. 8.3.

Smale obtained the remarkable result of giving a complete description of the dynamics of f on the set Λ by means of 'two-symbol' dynamics. In order to understand the essence of Smale's argument, let us follow a point of Λ under the iterates of f. At each iteration of f, the point is mapped onto one of *two* rectangular, disjoint pieces of $D \cap f(D) \cap \ldots f^n(D) \cap \ldots$

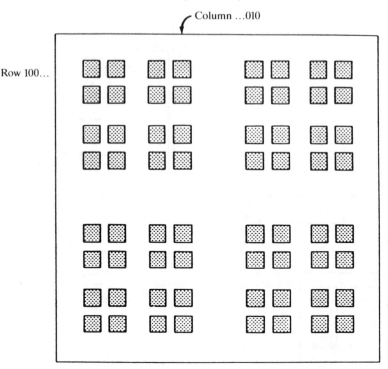

Fig. 8.3. A truncated, schematic picture of the invariant set of the horseshoe map.

Similarly, at each iteration of f^{-1}, the point is mapped onto one of two rectangular, disjoint pieces of $D \cap f^{-1}(D) \cap \ldots f^{-n}(D) \cap \ldots$. We can now choose the two symbols 0 and 1 and *uniquely* label those two pieces as '0' rectangle or '1' rectangle. Thus, corresponding to each point $x \in \Lambda$, we can uniquely define a bi-infinite sequence $s = \{\ldots, s_{-n}, \ldots, s_{-1}, s_0, s_1, \ldots, s_n, \ldots\}$, determined by the action of f on Λ, according to the rule

$$s_i = \begin{cases} 0 & \text{if } f^i \text{ maps to a '0' rectangle,} \\ 1 & \text{if } f^i \text{ maps to a '1' rectangle.} \end{cases}$$

Smale proved the following results:

(i)   there is a well-defined map ϕ which associates to each point $p \in \Lambda$ a bi-infinite sequence of the two symbols 0 and 1, in fact $\phi : \Lambda \to \Sigma^2$ is a homeomorphism;

(ii)   there is a direct relationship between the action of the map f on a point $x \in \Lambda$ and the action of σ on a sequence $s \in \Sigma^2$ (i.e., a point in the space Σ^2).

This can be grasped intuitively by observing that the (unique) sequence associated with $f^k(x)$ is obtained from the sequence associated with x merely by shifting the 'decimal point' in the former, n places to the right if $n > 0$, or n places to the left if $n < 0$. Thus, if $\phi(x)$ is the (uniquely defined) sequence associated with x, the sequence associated with $f^k(x)$ is the (uniquely defined) sequence $\sigma^k[\phi(x)]$, where σ^k is the kth iterate of the shift map defined above.

It follows that there exists a topological equivalence between f and σ, and consequently the dynamic behaviour of f on Λ exhibits all the complicated structures described in Propositions 8.1 and 8.1'. In particular, it possesses SDIC and therefore it is chaotic.

Furthermore, Smale showed that the dynamics on the 'horseshoe' set are structurally stable, in the sense that any map $\tilde{f}$ sufficiently close to f has an invariant Cantor set $\tilde{\Lambda}$ such that the dynamics of $\tilde{f}$ on $\tilde{\Lambda}$ are topologically equivalent to those of f on Λ.

The dynamics of Smale's map do not depend on the specific form of the map, but on the geometric properties of its action, the key elements of which are the following:

(i)   the set D is contracted, expanded and folded so that disjoint regions of it are mapped onto themselves;

(ii)   there are strong expansions and contractions in complementary directions.

A rigorous generalization of properties C.1 and C.2 above has recently been provided by Wiggins (1988, pp. 108–70), who defined the sufficient conditions (too complicated to be discussed here in detail) for a map to possess a chaotic invariant set.

Wiggins's criteria are very general in the sense that:

(i)   they extend to n-dimensional invertible maps ($n \geq 2$);

(ii)   they allow for subshifts of finite type as well as full shifts;

(iii) they extend to the case in which the invariant set of the map is nonhyperbolic. In this case, however, the notion of 'chaotic dynamics' must be somewhat adjusted.[8]

We have seen that the shift map on the bi-infinite sequence of finite

---

[8] One would say that – under the prescribed conditions – a map f possesses an invariant set of surfaces Λ such that there exist

(i)   a countable infinity of periodic surfaces in Λ;

symbols Σ^N is invertible. However, in certain applications, we want to investigate the behaviour of *non-invertible* maps, an outstanding example being the class of unimodal maps, which we shall discuss in Chapter 9.

In order to apply the method of symbolic dynamics to this case,[9] we need a space of simply-infinite sequences of two symbols, defined as

$$\hat{\Sigma}^2 \equiv \{s = (s_0 s_1 s_2 \ldots) \mid s_j = 0 \text{ or } 1\}.$$

The shift map $\sigma : \hat{\Sigma}^2 \to \hat{\Sigma}^2$ is given then by

$$\sigma(s_0 s_1 s_2 \ldots) = s_1 s_2 s_3 \ldots$$

In other words, the shift map 'forgets' the first entry in a sequence and shifts all other entries one place to the left. Since the first entry may be a 0 or a 1, clearly σ is a two-to-one map of $\hat{\Sigma}^2$, as is the case for a unimodal map f acting on the interval $I = [0, 1]$.

The shift map on $\hat{\Sigma}^2$ can be shown to be chaotic in the sense defined by Propositions 8.1 and 8.1'. Moreover, under certain conditions on the (unique) parameter characterizing the map f, there exists a subset $\Lambda \subset I$ which is a Cantor set, and such that the action f on Λ is topologically equivalent to the action of σ on $\hat{\Sigma}^2$. That is to say, under the prescribed conditions, f is chaotic on Λ.

(ii)   an uncountable infinity of nonperiodic surfaces in Λ;
(iii) a surface in Λ that at some point along its orbit is arbitrarily close to every other surface in Λ.

The dynamics of f in directions normal to surfaces in Λ can be said to be 'chaotic' in the same sense as in the case in which Λ is hyperbolic. However, the dynamics in directions tangent to the surfaces are unknown (Wiggins, 1988, pp. 159–61, th. 2.4.3).

[9] Cf. Devaney (1986, pp. 39–48).

9

Transition to chaos. Theoretical predictive criteria

9.1 INTRODUCTION

In the previous chapters, we have discussed a number of analytical and numerical criteria aimed at establishing whether certain types of irregular behaviour of deterministic dynamical systems can indeed be defined 'chaotic', and to provide quantitative characterizations of them. Thus, having located an apparently non-trivial attractor, we can inspect its geometrical properties and those of its Poincaré map, we can estimate its LCEs, evaluate its fractal dimension, study its power spectrum, and so forth.

The relevance of these procedures would be greatly enhanced if, in addition, we could describe the qualitative *changes* in the orbit structure of the system which take place when the control parameters are varied, certain bifurcations occur and complexity sets in. In this way, we would obtain not only a snapshot of chaotic dynamics, but also, so to speak, a 'stroboscopic' description of its emergence. Moreover, if we could provide a rigorous and exhaustive classification of the ways in which complex behaviour may appear, transition to chaos could be predicted theoretically, and potentially turbulent mechanisms could be detected in practical applications.

Unfortunately, the present state of the art does not permit us to define the prerequisites of chaotic behaviour with sufficient precision and generality. In order to forecast the appearance of chaos in a dynamical system, we are for the time being left with a limited number of theoretical predictive criteria and a list of certain typical (but by no means exclusive) 'routes to chaos'. This topic will be the subject of the present chapter.

Section 9.2 is a preliminary one and, strictly speaking, does not deal with transition to chaos. However, the concepts of 'horseshoe' and 'homoclinic orbits' are basic to the understanding of chaotic dynamics in general, and the presence of (non-attracting) horseshoes sometimes provides a valid predictive criterion for transition to chaos.

In Section 9.3 we deal with the generic 'codimension one routes to chaos', i.e., with those types of transition to chaos associated with (local or global) bifurcations that occur when a *single* control parameter is changed. Typically, there exist three such routes, namely:

(i)   period-doubling;
(ii) intermittency (or explosion);
(iii) saddle connection (or 'blue sky catastrophe').

The discussion of period-doubling will be put into the more general context of one-dimensional mappings, in view of the special attention that the latter have received in the economic literature.

In Section 9.4 we shall add a brief discussion of the so-called 'quasiperiodic route to chaos'. In so doing, we shall touch upon an important class of codimension two bifurcations, i.e., bifurcations occurring when *two* control parameters are simultaneously varied.

9.2 HOMOCLINIC ORBITS AND HORSESHOES

A quite general relation between homoclinic and horseshoe-like[1] invariant sets is established by the so-called Smale–Birkoff homoclinic theorem. The latter states that the presence of transversal homoclinic orbits (transversal intersections of stable and unstable manifolds of a hyperbolic fixed point) of a diffeomorphism implies the presence of horseshoe-like invariant sets.[2]

Although we do not have any similar general criterion to establish necessary and sufficient conditions for the occurrence of invariant, chaotic sets in systems of differential equations, it has become apparent that homoclinic and heteroclinic orbits often play an essential role in producing chaotic dynamics. The investigation of the orbit structure of a continuous-time system near homoclinic orbits is indeed one of the very few analytical instruments at present available to detect chaos in ordinary differential equations.

---

[1] Notice that the term 'horseshoe' is sometimes referred to as a (chaotic) invariant set, rather than to the map that generates it. The context should make it clear which of the two usages is employed.

[2] See Guckenheimer and Holmes (1983, p. 252). Another important result is found in Katok (1980), who proves that, for certain classes of Poincaré maps, the presence of positive entropy (in the sense discussed in Chapter 2) implies transversal homoclinic intersections.

For this purpose, the first step is to replace the dynamical system involving continuous time with a discrete-time system. We have already discussed how this may be done in the chapter devoted to Poincaré maps (see Chapter 4, where it was concluded that, under certain conditions, such a map can indeed be defined near a hyperbolic fixed point and near homoclinic orbits emanating from it.

The next step is to show the map possesses an invariant Cantor set on which it is topologically equivalent to a full shift (or to a subshift of finite type) on n symbols, along the lines discussed in Chapter 8. Although, as we have seen, general criteria exist to establish the latter properties, their essentially geometric nature makes it very difficult to apply them to systems which arise in practice, and very much is left to the investigator's ingenuity.[3]

There exists, however, one general result which we would like to mention. This is known as the 'Šilnikov phenomenon' and has been detected in a large number of applications, one of which will be discussed in Chapter 13. By means of a technique similar to the one just mentioned, Šilnikov (1965) proved the existence of horseshoe-like dynamics in a three-dimensional, continuous-time dynamical system in the presence of a homoclinic orbit (see Fig. 9.1). Šilnikov's argument can be summarized as follows.

Consider the system of differential equations

$$\dot{x} = f(x), \qquad x \in \mathbb{R}^3, \tag{9.1}$$

which has an equilibrium point at the origin, i.e., $f(0) = 0$, and whose Jacobian matrix, calculated at $x = 0$, has a real positive eigenvalue λ and a pair of complex conjugate eigenvalues $\sigma \pm i\theta$, $\sigma < 0$.[4] Let us *assume* that the system possesses a homoclinic orbit γ which tends to the origin as $t \to +\infty$, as well as when $t \to -\infty$. Then, the following theorem can be proved (see Guckenheimer and Holmes, 1983, p. 319):

Theorem 9.1. *If $|\sigma| < \lambda$, then the flow ϕ_t associated with system (9.1) can be perturbed to ϕ'_t such that ϕ'_t has a homoclinic orbit γ' near γ and the return map of γ' for ϕ'_t has a countable set of horseshoes.*

Expressed simply, this means that systems for which the stated conditions hold possess invariant chaotic sets. The same author has subsequently produced an extension of the above result to the case of arbitrary dimension (see Šilnikov, 1970).

---

[3] A well-known example in which the technique in question has been successfully applied is provided by Sparrow's study of the Lorenz model (see Sparrow, 1982).

[4] Analogous results can be found in the case in which $\lambda < 0$ and $\sigma > 0$.

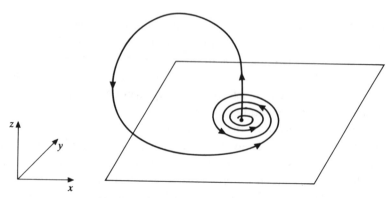

Fig. 9.1. The homoclinic orbit in (9.1).

As we observed at the beginning of this chapter, horseshoes are in-variant chaotic sets, but need not be attractors; in fact, Smale's own horseshoe is known to be non-attracting. Other possible horseshoe scenarios, characterized by chaotic attractors, could however be conceived by introducing appropriate conditions on the global dynamics of the systems.

At any rate, the detection of horseshoes is very helpful even when they are not attracting and cannot therefore be 'observed'. For one thing, their presence will have a striking effect on the transient dynamics of any system. In particular, in the neighbourhood of a horseshoe, orbits starting arbitrarily near one another will be exponentially separated in finite time and their fate may therefore be quite different. Secondly, the presence of a horseshoe may signal the impending appearance of a chaotic attractor, which may be triggered by one or another of the global bifurcations discussed in this chapter.[5]

We shall conclude this section by a brief remark on the delicate question of the detection of homoclinic orbits, whose (assumed) presence plays a fundamental role in the investigation of horseshoes.

Since homoclinic (and heteroclinic) orbits are global phenomena, there exist rather few analytical tools to prove their existence in given systems. A variety of perturbation techniques are available but the range of their application is rather limited. Most of these techniques derive from the early work of Melnikov (1963), and have been applied to problems where the dissipation is small (quasi-Hamiltonian), and for which the solutions

[5] See on this point, Thompson and Stewart, (1986, p. 277).

of the unperturbed equations are well understood.[6] Fortunately, it is sometimes possible to determine the existence of homoclinic orbits (of flows or maps) numerically, although of course this requires experience and judgement.

9.3 CODIMENSION ONE ROUTES TO CHAOS: ONE-DIMENSIONAL MAPS

Although the period-doubling route to chaos could be discussed in a rather general framework (cf. Eckmann, 1981, pp. 648–9), we shall deal with it in the context of one-dimensional maps, since they provide an interesting topic *per se*, and are by far the most common type of dynamical system encountered in economic applications of chaos theory.

One-dimensional models in economics, as well as in other social or natural sciences, are often too trivial or too specific to be truly relevant, and few results derived from them can be extended to more realistic, higher dimensional models. On the other hand, the study of iteration of maps of the interval has played a fundamental role in the recent growth of interest in, and understanding of, chaos theory. Such maps are simple enough to permit the application of certain analytical tools and lend themselves to quick and straightforward numerical computations. They are also reasonably easy to teach. Moreover, many dynamical systems which have been considered in applications are dissipative and possess attractors which, owing to a strong contraction of phase space volume, can be characterized by one-dimensional mappings with sufficient accuracy. For example, in the application discussed in Chapter 11 below, even though the phase space is ten-dimensional, the chaotic attractor of the system 'lives' in a three-dimensional space and, as we shall see, it can be effectively characterized by means of a one-dimensional map.

Finally, as far as economic applications are concerned, the vast majority of chaotic systems investigated consist of a single, finite difference equation, with a quadratic nonlinearity. For all these reasons, it will be helpful to discuss one-dimensional mappings in some detail. However, since there already exist several excellent treatments of the matter of various degrees of sophistication,[7] we shall present here only some essential

---

[6] For a thorough discussion of this question and a generalization of Melnikov's results, see Wiggins (1988, ch. 4).

[7] The literature on this subject is by now considerable. Rather than burdening our bibliography with a very long list of works, we shall mention here two very thorough treatments with rich bibliographies, namely, Collet and Eckmann (1980) and Devaney (1986).

results, concentrating on those issues which are relevant to the problem of strange attractors.

A one-dimensional map can be written as

$$x \mapsto f(x), \qquad x \in \mathbb{R} \tag{9.2}$$

or, in iterative form, as

$$x_{n+1} = f(x_n) \quad x \in \mathbb{R}, \quad n = 0, 1, 2, 3, \ldots, \tag{9.3}$$

where $f : \mathbb{R} \to \mathbb{R}$ is a continuous, smooth function. The literature on one-dimensional maps has concentrated on one-parameter families of maps $f : I \to I$, $I = [0, 1]$, belonging to the *unimodal* class, i.e., such that: (i) $f(0) = f(1) = 0$; (ii) f has a unique critical point[8] c with $0 < c < 1$.

A prototypical example of such families is the so-called logistic map

$$f_r(x) = rx(1 - x), \qquad 0 < r \le 4, \tag{9.4}$$

with the associated dynamical system

$$x_{n+1} = rx_n(1 - x_n) \quad n = 1, 2, 3, \ldots, \quad r < 0 \le 4. \tag{9.5}$$

The logistic map was employed as a model of population dynamics in Robert May's pioneering investigation (1976),[9] and has since inspired a large number of works by mathematicians as well as by practitioners of different disciplines, including mathematical economics.[10]

Unimodal maps are quite well understood and we shall list here some fundamental results which are independent of the specific form they take in different applications.[11]

1. Suppose f is three times differentiable, and either of the following equivalent properties holds:

(i) $\sqrt{|f'|}$ is a convex function on $x < 0$ and $x > 0$;

(ii)

$$S(f) \equiv \frac{f'''(x)}{f'(x)} - \frac{3}{2} \left(\frac{f''(x)}{f'(x)} \right)^2 < 0$$

($S(f)$ is usually called a *Schwarzian derivative*).

Then it has been demonstrated that:

(i)   for any given value of the parameter r, orbits starting from almost

---

[8] A point x is said to be a 'critical point' if $f'(x) = 0$. The critical point is, respectively, non-degenerate or degenerate, according to whether $f''(x) \neq 0$, or $f''(x) = 0$.

[9] An early discussion of a unimodal map can be found in Ulam and von Neumann (1947).

[10] For a (non-exhaustive) list of applications to economics, see Chapter 11 below.

[11] Some of these results in fact apply to a wider class of one-dimensional mappings.

all initial points on the interval $[0, 1]$ have the same asymptotic behaviour ('almost' here means 'excluding a set of Lebesgue measure 0', therefore excluding a set observable with zero probability). The choice of initial conditions is therefore not important as far as asymptotic behaviour of the map is concerned;

(ii)   the map can have *at most* one *stable* periodic orbit and, moreover, the orbit on the unique critical point is asymptotic to this stable orbit. The first part of this result (uniqueness of the stable periodic orbit) is important when we face the problem of distinguishing truly chaotic from periodic behaviour. For unimodal maps it could never be true that the limit set[12] includes a multiplicity of disconnected stable periodic orbits, each with a very small basin of attraction. Were this the case, small perturbations, physical or numerical, would prevent orbits from achieving their (periodic) asymptotic behaviour, suggesting the presence of chaos where there is none. Some authors have hinted that this might be the case with the (two-dimensional) Hénon map.

(iii)   When the limit set is not a periodic orbit, it can be *aperiodic but not chaotic*, or *chaotic*. In the former case, it is a Cantor set (it has non-integer fractal dimension), and the critical point lies in it. The limit set is an infinite one, but there is no sensitive dependence on initial conditions. The single Lyapunov characteristic exponent is equal to zero and orbits of two close initial points remain close. As we shall see, the aperiodic nonchaotic case occurs, for example, as a limit of a sequence of period-doubling flip bifurcations, but this is not the only possibility. In the chaotic case, the single LCE will be positive and there will be SDIC (nearby orbits diverge). This occurs, for example, in the quadratic map

$$x_{n+1} = rx_n(1 - x_n), \qquad r = 4.$$

(iv)   At present, it is not generally possible to detect the values of the parameter r for which the asymptotic motion is periodic, aperiodic or chaotic. All we can say is that the parameter range over which those types of behaviour occur is partitioned into subsets whose Lebesgue measures are all positive. In particular, we know that the set of all the parameter values which give birth to chaotic orbits possesses a positive measure on the r-axis, although those values

---

12 We do not use the phrase 'attracting set' in this context for reasons explained in Remark 9.1, below.

nowhere form an interval. The question may be raised, therefore, as to whether a phenomenon (in this case, chaos) which occurs on a parameter set of positive Lebesgue measures need be practically relevant. In fact, in the case under scrutiny, the set of parameter values for which the motion of the system is *not* chaotic is an open-dense set of the real line. Therefore, nonchaotic motion is *generic* in the sense usually given to this term. (Cf. Guckenheimer and Holmes, 1983, p. 258).

2. Let $f : \mathbb{R} \to \mathbb{R}$ be continuous. Then consider the following ordering of the natural numbers (the expression $p > q$ simply means 'p is listed before q'):

$$3 > 5 > 7 > \ldots > 2 \cdot 3 > 2 \cdot 5 > \ldots > 2^2 \cdot 3 > 2^2 \cdot 5 > \ldots$$
$$> 2^3 \cdot 3 > 2^3 \cdot 5 > \ldots > 2^3 > 2^2 > 2 > 1.$$

(In other words, we have first all the odd numbers; then 2 times the odd numbers, 2^2 times the odd numbers, 2^3 times the odd numbers, etc. There will remain only the powers of two which are listed in decreasing order.)

Then, a celebrated theorem of Sarkowskii (1964) states that:

Theorem 9.2.   *If f has a periodic point of* prime *period k, and $k > l$ in the above ordering, then f also has a periodic point of period l.*

Sarkowskii's theorem is very general as it does not apply only to unimodal maps but to all continuous maps of $\mathbb{R}$. On the other hand, it is a strictly one-dimensional result and it does not hold for higher dimensional maps, in fact not even for maps on the circle.

Within its limits, this theorem is a very powerful (and aesthetically beautiful) tool of investigation. Two corollaries are particularly interesting for us here. The first states that, if f has a periodic point whose period is *not* a power of 2, then f must have infinitely many period points. Conversely, finitely many periodic orbits must necessarily be powers of 2.

The second corollary has (misleadingly) become known as 'period three implies chaos' (cf. Li and Yorke, 1975). The truth in this phrase is that (i) period three is the period which appear first in Sarkowskii's ordering and therefore its presence implies the existence of all (infinitely many) other periods; (ii) period three implies the presence of an invariant set which is chaotic in the horseshoe sense.[13] This, in turn, implies very complex transients. What is not true, nor was it ever claimed by the authors of the phrase, is that period three implies a *chaotic attractor*. This may or may not be the case, but it is perfectly possible to have period three and a simple periodic attractor. In fact, the period-three window of a logistic map does not appear chaotic, since the chaotic set associated with it is not stable and therefore is not observable. More precisely, we could say that, in the case under scrutiny, m-almost all points on the interval $[0, 1]$ are attracted to the single, period-three orbit (m being the Lebesgue measure).

An n-dimensional generalization of the Li and Yorke theorem has been provided by Marotto (1978). Consider a map

$$x_{n+1} = G(x_n), \qquad x \in \mathbb{R}^n, \qquad (9.6)$$

where G is assumed to be differentiable in some closed ball $B_r(z)$ of radius r about z. Suppose $G(z) = z$ and the matrix $D_xG(x)$ has eigenvalues larger than one in modulus for all $x \in B_r(z)$. The point z is then a repelling fixed point. Moreover, suppose there exists a point $x_0 \in B_r(z)$, with $x_0 \neq z$, and a positive integer m such that $G^m(x_0) = z$ and $|D_xG^m(x_0)| \neq 0$. Then z is called a *snap-back repellor* and it can be proved that snap-back repellors imply (transient) chaos in the sense of Li and Yorke. In fact if $n = 1$, the existence of a snap-back repellor implies 'period three'. Another generalization has been provided by Diamond (1976). Let M be a set in $\mathbb{R}^n$ and $G: M \to \mathbb{R}^n$ be continuous. Then it can be proved that, if a non-empty compact subset X of M exists such that

$$X \cup G(X) \subset G^2(X) \subset M,$$

then the conditions for the Li and Yorke (transient) chaos obtain.

---

[13] More specifically, the relevant part of the Li and Yorke Theorem (1975, p. 246) can be described thus. Let $f: I \to I$, $I = [0, 1]$ be a continuous map. Let $\Lambda \subset I$ be an uncountable subset containing no periodic points. Then if the map f has a periodic point in I having period three, then:

(1) for every $x, y \in \Lambda$, $x \neq y$

$$\text{(i)} \qquad \limsup_{n \to \infty} |f^n(x) - f^n(y)| > 0,$$

and

$$\text{(ii)} \qquad \liminf_{n \to \infty} |f^n(x) - f^n(y)| = 0,$$

(2) for every point $x \in \Lambda$ and periodic point $y \in I$

$$\limsup_{n \to \infty} |f^n(x) - f^n(y)| > 0.$$

This implies that, in the aperiodic (chaotic) set Λ, orbits starting from two different (but arbitrarily close) points become infinitely often very close to one another and equally infinitely often finitely separated. Notice that 'period three' is a sufficient but not necessary condition for 'chaos' (in the sense of Li and Yorke) to occur.

Sarkowskii's theorem does not provide a complete classification of periodic orbits since it predicts only *the first appearance* of periodic orbits of different periods, while *repeated* periodic orbits of the same period, which occur for $k > 3$, remain undetected. A more complete characterization of periodic orbits (e.g., the exact number of periodic orbits of each given period k and the order of appearance of *stable* periodic orbits) may be obtained by using *symbolic dynamics* (and a new powerful version of it, known as *kneading theory*, which will not be not discussed here).

Remark 9.1. Although the aperiodic or chaotic limit sets which we have encountered in the study of the logistic map, can loosely be called 'attractors' in an operational sense, they are not attracting sets according to Definition 2.6. Consequently, they are not attractors according to Definition 2.7.[14] Indeed, in those cases, the dynamical systems described by the map are not dissipative. This is signalled by the fact that the (single) Lyapunov characteristic exponent is either zero (aperiodic case) or positive (chaotic case). In this respect, it is interesting to observe that, in general, the rate of growth of an n-dimensional volume of initial conditions is given by the sum of the largest n Lyapunov characteristic exponents. That sum will be zero for conservative systems and negative for dissipative systems. However, for one-dimensional *non-invertible* maps such as the logistic one, the link between the LCEs and the growth of volume is broken and we can have a bounded motion even in the presence of a (single) positive LCE. As we have seen in our discussion of the '1D map approach' in Chapter 4 non-invertible one-dimensional maps with a positive LCE can in certain cases be viewed as one-dimensional representations of higher dimensional, dissipative systems. An example of this approach will be discussed in Chapter 11 below.

9.4 PERIOD-DOUBLING

In the previous section, we have discussed the asymptotic behaviour of certain one-dimensional mappings with a special emphasis on the problem of the existence of chaotic behaviour. Here we shall address the question of how the behaviour of the map changes qualitatively when the control parameter is varied, and in particular how it becomes chaotic.

---

[14] Of course, it is perfectly possible to adopt a definition of 'attractor' which does not require the property of being an 'attracting set'. See, for example, Ruelle (1989, p. 25).

We shall try to answer this question in a rather simple manner, mainly relying on intuitive and geometrical considerations and referring the reader to the specialized literature for a more rigorous treatment.[15]

To fix ideas and to make reference to well-known applications easier, we shall discuss the question in relation to the logistic map[16]

$$x_{n+1} = rx_n(1 - x_n). \tag{9.7}$$

Let us now consider how the behaviour of (9.7) changes when r is varied. There exist two fixed points, namely $\bar{x}_1 = 0$ and $\bar{x}_2 = 1 - 1/r$ and their eigenvalues are respectively $\lambda = r$ and $\lambda = 2 - r$. Thus, for $0 < r < 1$ the origin of the co-ordinates is stable and the second (negative) fixed point unstable.

At $r = 1$ we have a *transcritical bifurcation*, with exchange of stability: the two fixed points merge and, past $r = 1$, $\bar{x}_1$ becomes unstable and $\bar{x}_2 > 0$ is now stable. At $r = 3$, the stable branch loses stability by a *flip bifurcation*: the two fixed points are now both unstable, and a stable 2-cycle (p, q) is born, where

$$q = rp(1 - p), \qquad p = rq(1 - q).$$

Consequently, p and q are the roots of

$$r^2x^2 - r(r + 1)x + (r + 1) = 0,$$

and their multipliers are solutions of

$$\lambda = -r^2 + 2r + 4.$$

Simple calculations will show that the initially stable 2-cycle loses stability at $r = 1 + \sqrt{6} \approx 3.449490$, where λ goes through -1 (a flip bifurcation), and a new, stable 4-cycle is created. This occurrence is repeated over and over again, and, if completed, leads to an infinite sequence of flip bifurcations[17] and *period-doublings*.[18] The sequence $\{r_k\}$ of values of r at which k-cycles appear has a finite accumulation point $r_\infty \approx 3.569446$, involving an infinity of periodic orbits.

---

[15] Besides the cited works by Devaney (1986) and Collet and Eckmann (1980), see, for example, Holmes and Whitley (1983); Guckenheimer and Holmes (1983, pp. 306–11).

[16] It will have been noticed that (9.7) maps $[0, 1]$ into $[0, 1]$ only for $0 < r \le 4$. Later we shall briefly mention the case of $r > 4$.

[17] Notice a possible source of confusion here. Some authors state that period-doubling occurs via *flip* bifurcations (e.g., Lauwerier, 1986, pp. 43–4), others via *pitchfork* bifurcations (e.g., Schuster, 1989, pp. 42–3). They are both right, but the former refer to bifurcations of the stable branches of the original map $f(x)$, while the latter refer to bifurcations of fixed points of the iterated map $f^k(x)$.

[18] Notice that the period-doubling cascade need not be complete. There may be only a finite number of doublings followed by 'underdoublings' or by other bifurcations.

Feigenbaum (1978) discovered, and Collet *et al.* (1980) and Landford (1982) later rigorously proved, that there are two universal quantitative features in this scenario, namely:

(i)   The convergence of r to r_∞ is controlled by the universal parameter $\delta \approx 4.669202$. If we call Δ_i and Δ_{i+1}, respectively, the distances on the real line between successive flips (i.e., $\Delta_i = r_k - r_{k-1}$ and $\Delta_{i+1} = r_{k+1} - r_k$), we have:

$$\lim_{i\to\infty} \frac{\Delta_i}{\Delta_{i+1}} = \delta.$$

The number δ has become known as the 'Feigenbaum constant' (see Fig. 9.2b).

(ii)   The relative scale of successive *branch splittings* (see Fig. 9.2a) is controlled by the universal parameter $\alpha \approx 2.502907$. One way of representing the convergence to α is the following (see Schuster, 1989, pp. 44–6). Let d_n be the distances of the fixed points closest to $x = 1/2$ ($= \max_x f(x)$) for superstable[19] 2^n-cycles. Then,

$$\lim_{n\to\infty} \frac{d_n}{d_{n+1}} = -\alpha.$$

It is interesting to notice that the quantitative properties of branch splitting are related to those of the power spectrum of the map. As more and more k-cycles are born, additional frequency components appear in the spectrum. The relative heights (log of power) of successive subharmonics drop, on the average, at a rate which depends on α.[20]

The parameters δ and α are universal in the sense that, for a very large class of maps, they have the same values. These numbers have also been measured in a variety of physical experiments. There is some evidence, moreover, that this universality applies not only to simple one-dimensional maps, but also to continuous, multi-dimensional systems which, owing to strong dissipation, can be characterized by a one-dimensional map.[21] For practical purposes, universality often permits

---

[19] A superstable k-cycle of a map $f(x)$ with elements $\{x_0, x_1, \ldots, x_{k-1}\}$ is defined by

$$\frac{df^k(x_0)}{dx} = \prod_i f'(x_i) = 0, \qquad i = 0, 1, \ldots, k-1,$$

which implies that, in the case of a unimodal map, a superstable cycle always contains the critical point because this is the only point at which $f'(x) = 0$.

[20] Cf. Cvitanovic (1984, pp. 22–3).

[21] Cf. Franceschini and Tebaldi (1979), in which infinite sequences of period-doublings have

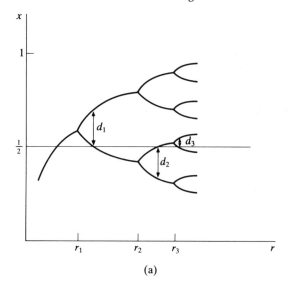

(a)

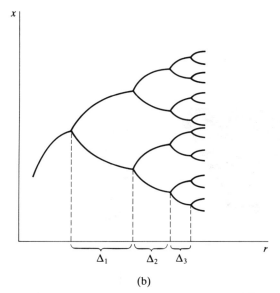

(b)

Fig. 9.2. The universal constants of the period-doubling scenario.

been found in a system of five differential equations. The numerically-calculated values of δ and α agree with the Feigenbaum constants. Evidence of period-doubling sequences in multidimensional, continuous-time systems will also be found in Chapters 11 and 13 below.

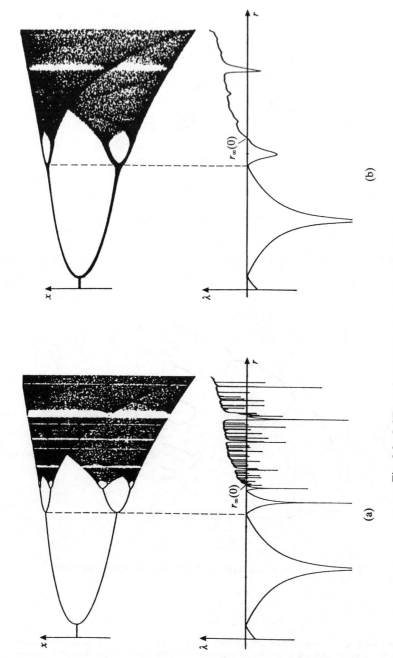

Fig. 9.3. LCEs and periodic windows in the chaotic zone.

us to make accurate predictions of r_∞ as soon as the first few flips are known. For example, in the case of the logistic map, the knowledge of r_2 and r_3 allows us (using a pocket calculator) to predict r_∞ correctly to four decimal places.

At $r = r_\infty$ the motion of the system is aperiodic, but not chaotic in the sense discussed above. The limit set (the so-called Feigenbaum attractor) is a Cantor set with fractal dimension ≈ 0.538, a LCE equal to zero, and consequently no SDIC. Just past r_∞, we enter what is usually called the 'chaotic regime'.

The chaotic regime ($r_\infty < r \le 4$) is characterized by the following properties:

(i) the chaotic intervals (i.e., the intervals of values of x within which the motion is chaotic) move together by inverse bifurcations until the iterates become distributed over the whole interval $[0, 1]$ at $r = 4$;

(ii) there is an infinite number of small windows of r-values for which there exist stable k-cycles. First to appear are those of even periods. Next, odd cycles appear in descending order. The period 3 cycle has a particularly wide window ($3.828427 \le r \le 3.841499$). Inside the windows the asymptotic motion is not chaotic, the LCE is negative and there is no SDIC.

(iii) Outside the windows, there are no stable periodic orbits, but there is an infinite number of unstable ones. Although there exist infinitely many values of r for which the dynamics are chaotic, they are everywhere densely interwoven with nonchaotic values of r, and nowhere do they form an interval.

In fact, if we look at the diagram showing the values of the LCEs corresponding to different values of r (Fig. 9.3a), we can see that, for $r_\infty < r \le 4$, although LCEs look mostly positive (whence the appellation 'chaotic zone'), the diagram shows an infinity of downward spikes.

In applications, however, where the presence of a certain amount of noise must always be assumed, the relevance of almost all periodic windows may be questioned. As can be seen in Fig. 9.3b, the addition of noise washes out the finer structure in the bifurcation and the LCE diagrams, still leaving a sharp transition to chaos, but only the largest periodic windows.[22]

The limit case $r = 4$ deserves special attention since it can be proved

---

[22] Cf. Crutchfield, Farmer and Huberman (1982); also Schuster (1989, pp. 59–62).

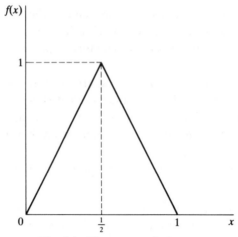

Fig. 9.4. The symmetrical tent map.

analytically that, for this particular value of r, the logistic map is in fact chaotic. First of all, note that the LCE of a map is invariant w.r.t. an invertible change of co-ordinates (Oseledec, 1968). Suppose now we make the following transformation:

$$y = \left(\frac{2}{\pi}\right) \sin^{-1} \sqrt{x}.$$

Then the map $f(x) = 4x(1-x)$ can be transformed into the symmetric, semilinear 'tent' map $\tilde{f}: I \to I$, $I = [0,1]$ (see Fig. 9.4)

$$\tilde{f}(y) = \begin{cases} 2y, & \text{for } 0 < y < 1/2, \\ 2 - 2y, & \text{for } 1/2 < y \le 1. \end{cases} \tag{9.8}$$

We can now evaluate the LCE exactly, thus:

$$\chi = \lim_{n \to \infty} \frac{1}{n} \sum_{i=1}^{n} \ln \left| \frac{d\tilde{f}}{dy_i} \right|,$$

where $\{y_1, y_2, \ldots, y_n\}$ is the sequence of values of y obtained by iterating the map $\tilde{f}$. Hence, we have

$$\chi = \ln 2 > 0.$$

From this we conclude that $\tilde{f}$ possesses SDIC and, by virtue of the Oseledec theorem, so does f.

We shall conclude this section with the following remarks.

Remark 9.2.   The route to chaos we have been discussing is probably the most common one in applications. Here the asymptotic orbit structure evolves gradually through a sequence of continuous, 'safe boundary' bifurcations, brought about by slow variations of the control parameter. Consequently, there are plenty of forewarnings of the impending chaotic behaviour. As we shall shortly see, however, the smooth, period-doubling route to chaos is not the only one present in one-dimensional mappings of the logistic type.

Remark 9.3.　　Within the *period 3* window, we have a somewhat intermediate case in which there are three narrow bands which are visited with regular periodicity, but within each of which the motion is chaotic. This occurrence has been labelled by Lorenz (1980) *noisy periodicity.*

Remark 9.4.　　For $r > 4$, most orbits originating in the interval $[0, 1]$ will eventually escape to $-\infty$. However, it can be proved that for these values of r, the logistic map possesses an invariant chaotic set which is non-attracting.

Remark 9.5.　　The fact that, in the chaotic region, chaotic and non-chaotic parameter values are densely interwoven gives rise to a *sensitive dependence on parameters* (SDP), i.e., arbitrarily small changes in the parameter values lead to drastic alterations in the mode of behaviour of the system. As Schuster (1989, p. 63) has aptly observed, the practical implications of SDP are even worse than those of sensitive dependence on initial conditions. When chaos and SDIC occur in a deterministic system, although we are unable to predict exactly the behaviour of the system, we can resort to statistical predictions. However, SDP makes statistical averages unstable under variations in parameters, since those averages may be quite different in two situations (chaotic and periodic), even though they are very close in the parameter space.

9.5　INTERMITTENCY

Generally speaking, a signal (the value of a variable changing in time) is said to be *intermittent* if it is subject to infrequent variations of large amplitude.

In the present context, we shall apply the term 'intermittency' to a phenomenon – common in low-dimensional dynamical systems depending on a single parameter – which can be described as follows.

For values of the control parameter r less[23] than a certain critical threshold r_c, the system has a stable limit cycle (a stable fixed point of

---

[23] Of course, the reasoning that follows could be restated inverting the relevant inequality signs.

the corresponding Poincaré map). Consequently, for $r < r_c$, the behaviour of the system is very regular, at least for initial conditions not too far from its stationary regime. When r becomes slightly greater than r_c, the oscillations continue but they are interrupted from time to time by a different, 'irregular' type of behaviour whose structural characteristics depend little on r, but whose *average frequency* does depend on it. In fact, if we start from $r > r_c$ and we reduce r continuously, the frequency is reduced gradually and tends to zero as $r \to r_c^+$. Vice versa, when we increase r past r_c, the 'irregular' behaviour becomes progressively more frequent until any memory of the regular oscillations is lost.

Thus, we can list the following abstract topological criteria characterizing intermittency:

(i)   a discontinuous (catastrophic), local bifurcation leading to the destruction of a 'small' attractor (e.g., a limit cycle corresponding to a fixed point of the Poincaré map) and to the creation of a 'larger' attractor which contains the locus of the former;

(ii) a continuous change of a certain measure density of orbits on the 'large' attractor. That is to say, after the bifurcation, the nearer the parameter to its critical value, the more likely it is for the system to be found near the locus where the 'small' attractor was before the bifurcation.

If we consider the problem in terms of local bifurcations of a map (which may or may not be the Poincaré map of a continuous system), we can distinguish between three different types of intermittencies, according to the different ways in which the 'small' attractor loses its stability, namely:

(i)   a saddle-node bifurcation;
(ii) a subcritical Hopf (or Neimark) bifurcation;
(iii) a subcritical flip bifurcation.

The destruction of a 'small' attractor by a local, catastrophic bifurcation does not necessarily imply transition to chaos. The 'large' attractor to which the system settles afterwards may be a nonchaotic (e.g., a quasiperiodic), or a chaotic one, according to the global structure of the phase space (and the corresponding transients) that prevails before the bifurcation.

We shall not discuss the various types of intermittency and the rather subtle questions associated with them in any detail.[24] Instead, in order

---

[24] For a more thorough treatment of the matter, see Schuster (1989, pp. 79–103) and Bergé, Pomeau and Vidal (1984, pp. 223–63).

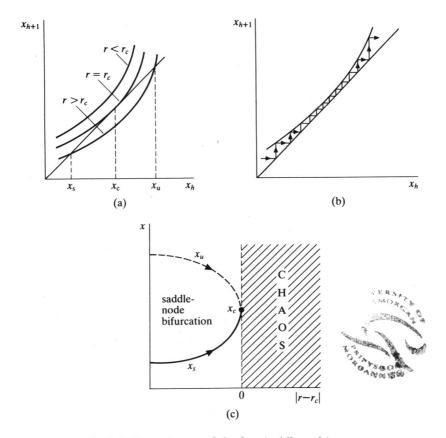

Fig. 9.5. Intermittency of the first (saddle-node) type.

to illustrate the rather abstract considerations developed so far, we shall give an intuitive description of a specific example of transition to chaos via the first type of intermittency, which occurs in the case of the logistic map[25]

$$x_{n+1} = f(x_n) = rx_n(1 - x_n). \tag{9.9}$$

The analysis developed in the preceding section indicates that, when $r_c \approx 1 + \sqrt{8}$ (well within the 'chaotic region'), map (9.9) has a period 3 cycle. Moving further inside the window, we have subsequent flip bifurcations, leading to period 3^n ($n = 2, 3, \ldots$) cycles.

The situation is schematically illustrated in Figs. 9.5 and 9.6. For values

[25] Cf. Schuster (1989, pp. 82–3).

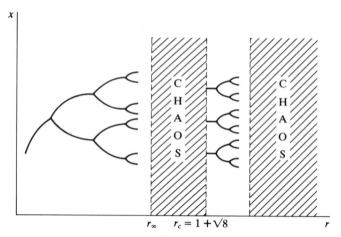

Fig. 9.6. Period 3 window in the chaotic zone.

of r slightly greater than r_c the motion is regular, 3-periodic. However, if we iterate map (9.9) for values of r slightly smaller than r_c, we note that a regular, almost periodic motion is interrupted from time to time by bursts of chaos. The iterates of the map accumulate in the neighbourhood of the 3-cycle points-to-be, and the average duration of regular dynamics is a continuous (inverse) function of the distance $|r_c - r|$.

The explanation of this rather curious fact can be found by considering the fixed point of the map $f^3(x_n) \equiv f(f(f(x_n)))$ corresponding to the 3-cycle (see Fig. 9.5a). At $r > r_c$, the map $f^3(x_n)$ has one stable fixed point; at $r < r_c$ it has none. The destruction of the fixed point takes place by a fold bifurcation at $r = r_c$, where the stable and the unstable fixed points of the map coalesce and disappear.

The dynamics of the map for $r < r_c$ can be observed in Fig. 9.5b. It can be seen that in the vicinity of the locus where the stable fixed point was, the motion of the system slows down as if it were still 'looking for it'. After a certain number of apparently regular iterations (the nearer r to r_c, the more numerous are the iterations), the system leaves the neighbourhood and wanders away in an irregular fashion until it is reinjected in the 'channel' between the curve of the map and the bisector, and so on.

Remark 9.6.   The fold bifurcation which takes place at $r = r_c$ is the only mechanism through which, in the logistic and similar maps, an odd number of fixed points (an odd-period orbit) could be created.

Remark 9.7. Intermittency transition to chaos is also called *interior catastrophe* (or *interior crisis*), making reference to the fact that the locus of the old attractor is included in the 'new' one.

Remark 9.8. There can be an intermittent transition from a 'smaller' chaotic attractor to a 'larger' one, containing the locus of the former. An example of this occurrence can be observed for the logistic map at the end of the period 3 window.

Intermittency of the type we have described, has been also found in continuous-time systems like the one represented by the Lorenz model.[26] The transition from a 'small' to a 'large' chaotic attractor, of which we have found some numerical evidence in the system discussed in Chapter 13 below, could also have been generated by intermittency.

9.6 SADDLE CONNECTION, OR 'BLUE SKY CATASTROPHE'

The phenomenon of intermittency occurs through a local/global bifurcation, i.e., a (discontinuous) change in the local stability properties of a fixed point, accompanied by global changes in the phase portrait of the system.

On the contrary, 'blue sky catastrophe' is due to an entirely global bifurcation, which leaves unchanged the properties of the existing fixed points. This rather colourful term was coined by Ralph Abraham,[27] and refers to a situation in which, when a single control parameter is varied, an attractor of the system suddenly disappears 'into the blue', and the orbits originating in what used to be the basin of attraction of the destroyed attractor wander away, and possibly move towards another remote attractor. A simple example of a blue sky catastrophe is depicted by Fig. 9.7 which is derived from the study of a Van der Pol system of two ordinary, scalar differential equations,[28] namely:

$$\dot{x} = ky + \mu x(b - y^2),$$
$$\dot{y} = -x + C.$$

For values of C smaller than a critical value $\bar{C}$, there exists an (unstable) saddle point and a stable limit cycle. As we increase C, the limit cycle moves progressively nearer the saddle point until, at $C = \bar{C}$, they collide.

[26] Cf., for example, Bergé, Pomeau and Vidal (1984, pp. 241–3).

[27] Abraham and Marsden (1978).

[28] The description that follows is based on Thompson and Stewart (1986, pp. 268–72).

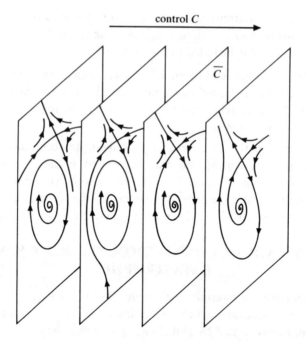

Fig. 9.7. A blue sky catastrophe for a two-dimensional flow.

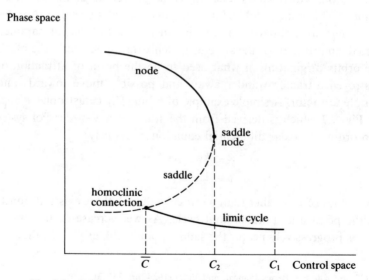

Fig. 9.8. A hysteresis loop involving a blue sky catastrophe.

At this point, a stable and an unstable branch of the saddle coincide with the location of the cycle, creating a homoclinic connection, and the cycle itself disappears. (The homoclinic connection can be thought of as an infinite-period cycle.)

Past $\bar{C}$, all the orbits originating in the region of the saddle point considered so far (below as well as above its stable branches), wander away and, if another remote attractor exists, they move towards it. Notice that, unlike the case of intermittency, this attractor does not contain the locus of the 'old' one, and may be far from it.

The blue sky catastrophe can be accompanied by another interesting phenomenon, hysteresis, which does not appear in intermittency. A simple case is illustrated by Fig. 9.8. At C_1, the only attractor is a stable limit cycle. If we reduce C continuously, at $\bar{C}$ the limit cycle touches the saddle point, a blue sky catastrophe occurs, the limit cycle is destroyed and the system moves up towards the remote stable node. If we reverse the change, however, and we *increase C* past $\bar{C}$, although a limit cycle is created, the node is not destroyed and keeps its stability until we reach C_2, at which point a fold catastrophe takes place, the node disappears and the system jumps back to the limit cycle. Thus, the changes in the global dynamics of the system are different according to whether the control parameter is increased or decreased. This is indeed the essence of hysteresis.

In systems of differential equations of dimension ≥ 3 (or in maps of any dimension), blue sky catastrophes may involve more complicated types of attractors, leading to the creation of chaotic attractors via collision with saddle-type objects. Even the simplest (three-dimensional) chaotic blue sky catastrophe is a very complex event which we cannot discuss here in detail. To give the reader an idea of the possible sequences of steps leading to such a global bifurcation, we shall reproduce and comment on the beautiful stroboscopic diagram of Abraham and Shaw (1988, p. 143), showing the catastrophic creation (destruction) of a horseshoe-like chaotic attractor.

Fig. 9.9 represents five Poincaré maps of a three-dimensional flow, each of them corresponding to a certain value of the control parameter. Reading from left to right we can identify (intuitively, not rigorously!) the following steps:

(1) There are two fixed points of the map (corresponding to cycles of the flow), a saddle and a repellor of a spiral type.

(2) As the control parameter increases, the stable and unstable mani-

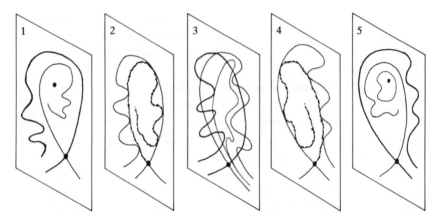

Fig. 9.9. A blue sky catastrophe for a three-dimensional flow (Poincaré map).

folds of the saddle approach one another. When they touch, generically they must have infinitely many points of tangency, corresponding to infinitely many folds of the invariant sets of the 3D flow. The contact is one-sided, the unstable manifold encircling the stable one. The transient diagram here is very complex due to the presence of horseshoe-like structures.

(3) For a still higher value of the parameter, a chaotic *attractor* appears, whose Poincaré map is a fractal collection of points lying on a curve with a characteristic shape, called by Abraham and Shaw a 'bagel'.

(4) This depicts a situation perfectly symmetrical with that shown by (2), with the stable manifold encircling the unstable one. For obvious reasons, the range of values of the parameter corresponding to (2)–(4) is called the 'tangle interval'.

(5) Beyond the tangle interval, tangencies of invariant sets have disappeared, and orbits starting 'inside' the region encircled by the stable manifold of the saddle, spiral towards an attracting invariant torus (the bulges of the torus are 'reminiscences' of the disappeared tangencies).[29]

Examples of chaotic blue sky catastrophes have been found in three-dimensional systems of differential equations (Rössler, 1976); in two-

---

[29] Notice that the illustrations above are necessarily incomplete and collections of points have been replaced by continuous curves for expediency.

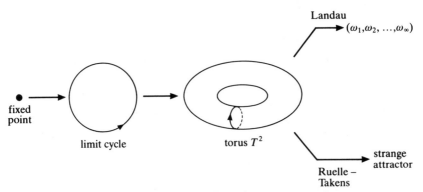

Fig. 9.10. Landau and Ruelle–Takens routes to chaos.

dimensional maps of the Hénon type (Simó, 1979); and even in one-dimensional maps of the logistic type (Grebogi *et al.*, 1983).

To conclude, notice that in the case of transition to chaos by global bifurcations of the blue sky catastrophe type, the appearance of a chaotic attractor is foreshadowed by the existence of horseshoe-like sets closely related, in theory and in practice, to transverse homoclinic orbits. Of course, when they are not attracting, these sets can be observed with probability zero, and their presence can be detected only in transients. For this reason, in the search for symptoms of impending chaos, it is very helpful to identify theoretically the existence of homoclinic trajectories and horseshoes, as discussed above in this chapter and in Chapter 8.

9.7 QUASIPERIODIC ROUTE TO CHAOS

The idea that quasiperiodicity is the fundamental intermediate step in the route to chaos, is a long-standing one. As early as 1944, the Russian physicist Landau suggested that turbulence in time, or chaos, is a state approached by a dynamical system through an infinite sequence of Hopf instabilities which take place when a certain parameter is changed. Thus, the dynamics of the system would be periodic after the first bifurcation, and quasiperiodic after successive bifurcations, with an ever-increasing degree of quasiperiodicity, leading in the limit to turbulent, chaotic behaviour. (See Fig. 9.10.)

The conjecture that Landau's is the only, or even the most likely route to chaos was rejected on the basis of two new basic results, namely:

(i) from a theoretical point of view, Ruelle and Takens (1971) and

Newhouse, Ruelle and Takens (1978) proved that systems with 4-torus (or even 3-torus) attractors are unlikely to be observed, as they are easily perturbed to chaos;[30]

(ii) Lorenz's (1963) work on turbulence showed that complexity (in the sense of a large dimension of the system) is not a necessary condition for chaos to occur, and that low-dimensional systems are perfectly capable of producing chaotic output.

Thus, although experimental results seem to suggest the possibility of a direct transition from quasiperiodicity to chaos, mathematically this is still an open question which we cannot discuss exhaustively here.[31] However, some of the mathematical aspects of the question have been investigated by studying the so-called circle map, which, in turn, is related to the Hopf (or Neimark) bifurcation of discrete-time dynamical systems. Since this type of bifurcation is interesting in itself, and its presence has been found in a number of economic applications, including the one discussed below in Chapter 12, we shall devote to it the next few paragraphs.

Let us now return for a moment to the discussion of bifurcations in Chapter 2 above. We were considering a *one-parameter* family of two-dimensional maps, such that, for a certain value of the controlling parameter there is a Hopf–Neimark bifurcation and – assuming away the case of 'strong resonance', i.e., the case in which $\lambda^i(\mu_c) = 1$, for $i = 1, 2, 3, 4$, – an invariant circle appears.[32] In order to investigate the dynamics of this family of maps after the Hopf bifurcation, we must embed it in a *two-parameter* family. This could be done, for example, by considering the path followed in the complex plane by the eigenvalue $\lambda(\mu)$ of the Jacobian matrix J_μ. Changing only one parameter would correspond to moving along an arc in the plane, as depicted in Fig. 9.11.

However, it can be shown that, in this case, by analogy with what happens to unimodal maps for the period-doubling phenomenon, the special functional form of the map is – within certain constraints –

---

[30] The results given by these authors are sometimes said to define a possible route from quasiperiodicity to chaos. However, as Thompson and Stewart observe (1986, pp. 287–8), those authors showed only that arbitrarily near (in the appropriate space) to a vector field having a quasiperiodic attractor, there exist vector fields having chaotic attractors. They did not prove that there exists an arc in control space, along which a transition from quasiperiodicity to chaos can be realized.

[31] The interested reader can consult, for example, Schuster (1989, pp. 145–85); Bergé, Pomeau and Vidal (1984, pp. 159–91) and the bibliography quoted therein.

[32] The 'strong resonance' case is difficult and not yet fully understood. The interested reader can consult Whitley (1983, pp. 205–9).

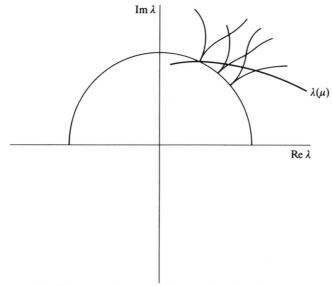

Fig. 9.11. Arnold tongues in the $(\text{Im }\lambda, \text{Re }\lambda)$ plane.

unimportant. In what follows, therefore, we shall study the simple case of the two-parameter family of the so-called circle maps[33]

$$\theta_{t+1} = f(\theta) = \theta_t + \Omega + \frac{K}{2\pi}\sin 2\pi\theta_t \quad \text{mod } 1, \qquad (9.10)$$

where θ represents the position along the circumference (so that $\theta = 0$ is identified with $\theta = 1$), and K is the (normalized) amplitude.

The various possibilities can be thus summarized, following Thompson and Stewart (1986, pp. 286–7):

(i)   For small K, i.e., near the bifurcation, the iterated circle map tends to a unique attracting fixed point if Ω is a rational number p/q, and the same is true for more or less large intervals of Ω near p/q. The periodic 'windows' are bounded by two arcs of saddle-node bifurcations. They have a characteristic shape called 'Arnold horns', or 'Arnold tongues', some of which are illustrated in

---

[33] The features of the map (9.10) relevant for the present discussion are:
  (i) $f(\theta + 1) = 1 + f(\theta)$;
  (ii) For $|K| < 1$, $f(\theta)$ is a diffeomorphism;
  (iii) At $|K| = 1$, $f^{-1}(\theta)$ becomes non-differentiable;
  (iv) For $|K| > 1$, no unique inverse of $f(\theta)$ exists.

(Cf. Schuster, 1989, p. 157.)

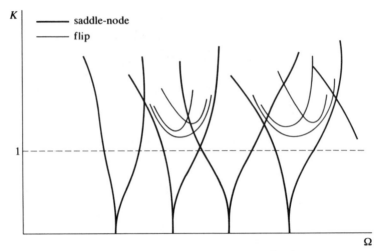

Fig. 9.12. Arnold tongues in the (K, Ω) plane.

Fig. 9.12. As K increases, and we move off the bifurcation point, the windows overlap one another, and the stronger the coupling, the more probable the mode-locking (i.e., synchronization of different frequencies) becomes.

(ii) $K = 1$ is the boundary between invertible and non-invertible circle maps. The non-mode-locked intervals between Arnold tongues form a Cantor set with zero measure.

(iii) For $K > 1$ the map $f(\theta)$ becomes non-invertible, chaos becomes possible, but chaotic and nonchaotic regions are closely interwoven on the (K, Ω) parameter space.

The question is now whether, moving in the parameter space, we can follow a 'route to chaos'. To answer this question, consider that, changing one parameter only corresponds to moving along an arc in the (K, Ω) plane.

In this case, it has been proved that mode-locking always occurs when K is increased beyond 1.[34] Then, the transition to chaos can only take place through one of the 'canonical' (codimension one) routes discussed above (i.e., period-doubling, intermittency and blue sky catastrophe).

When, on the contrary, *both* the parameters K and Ω can be controlled simultaneously, it is possible to reach a point on the line $K = 1$ belonging

---

[34] Cf. Bak, Bohr and Jensen (1984).

to the non-mode-locking Cantor set, without entering any of the Arnold tongues. The question is still open whether we can move beyond $K = 1$, going directly from quasiperiodic to chaotic behaviour and skipping all the mode-locking regions. As we have said, a fully developed mathematical answer to this question is still lacking, although experimental results seem to suggest that the transition is indeed possible.[35]

[35] See, for example, Franceschini and Tebaldi (1979).

10

Analysis of experimental signals – some theoretical problems

10.1 THE THEORETICAL PROBLEM

When investigating the dynamics of a real system, for example the evolution in time of a certain economy, we typically monitor one or a few scalar variables that change in time. Suppose we have access to GNP data evaluated at equally spaced sampling times and, to simplify matters, that there are no errors in measurement – i.e., we have a string of observations, $Y(t)$, $Y(t+\tau_s)$, $Y(t+2\tau_s)$, etc. We conjecture that the time series is the output of a dynamical system including a certain number of variables and obeying certain dynamical laws, and we would like to ascertain its basic qualitative features solely by analysing the single scalar signal $Y(t)$.

The enterprise is apparently hopeless, since we do not even know how many other variables play an essential role in determining the evolution of the economy, let alone have the ability to measure them. In mathematical terms, we do not even know the dimension of the system, which in principle can be indefinitely large. Two circumstances, however, make this seemingly impossible project feasible.

First of all, the *attractors* of large (or even infinite dimensional) systems may have low dimensions. Therefore, if we neglect transients (i.e., if we are only interested in the asymptotic behaviour of the system), the study of a low-dimensional geometrical object may provide all the information we need. Secondly, and most importantly, there exist theorems proving the possibility of extracting from a univariate time series the desired information about the system which, by hypothesis, has generated it (henceforth called the 'original system').

For this purpose we take n consecutive elements of the time series of

¶ Chapter 10 and the related Chapter 14 have been largely based on unpublished material by Roberto Perli Traverso. The author wishes to thank him for giving permission to use it in the preparation of this book.

Y, i.e.,

$$Y(t), Y(t + \tau_s), Y(t + 2\tau_s), \ldots, Y(t + (n-1)\tau_s), \qquad (10.1)$$

and we consider these elements as a vector defining a point in $\mathbb{R}^n$. If we follow the evolution in time of this vector – as described by the data of Y – we obtain an orbit in a space $B \subset \mathbb{R}^n$. Under rather mild conditions, it can be shown that, for almost any choice of the time delay τ_s, the dynamics in the 'pseudo phase space' B have the same (asymptotical) properties as those of the 'original system', provided the dimension n of vector (10.1) is sufficiently large relative to that of the true attractor.

More formally, the basic result can be stated as follows:[1]

Theorem 10.1.   *Let A be a compact manifold of dimension m. For pairs (F, v), where F is a smooth vector field and v a smooth function on A, it is a generic property that $\Phi_{F,v}(y): A \to \mathbb{R}^{2m+1}$, defined by*

$$\Phi_{F,v}(y) = (v(y), v(\phi_1(y)), \ldots, v(\phi_{2m}(y)))^T,$$

is an embedding,[2] *where ϕ_t is the flow of F, and ^{T} indicates transpose.*

In this context, the function $v(y)$ can be taken as a measurement made on the system at the point $y \in A$, i.e., in the case considered above, $v(\phi_i(y))$ would be equal to an observation of Y at time i.

Takens's theorem in principle guarantees that, if n is sufficiently large vis-à-vis the dimension of the manifold on which the attractor lies, the n-dimensional image of the (conjectured) attractor provides a correct topological picture of its dynamics. For example, periodic orbits (cycles) on the attractor will correspond to periodic orbits in the reconstructed phase space; chaotic orbits of the original system will look chaotic in that space, and so on. More specifically, an attractor obtained from the time series (10.1) will have *the same positive LCEs* as those of the corresponding attractor of the original system.

We notice that, in general, the choice of the delay τ_s is not critical, at least in principle. However, in practical calculations it is advisable to choose it small in relation to the average orbit excursion along the attractor.

The essential question is instead how large must n be, i.e., the dimension of the reconstructed phase space. To understand this, consider that the reconstruction defined above provides an n-dimensional projection of an

---

[1] This version of Takens's (1981) theorem is taken from Broomhead and King (1986b).

[2] Let M be a compact set in a Banach space B, and E a subspace of finite dimension. Then an *embedding* is a smooth map $\Phi : B \to E$, such that $\Phi(M) \subset E$ is a smooth submanifold of E and Φ is a diffeomorphism between M and $\Phi(M)$.

attractor which has a finite dimension but is immersed in a phase space whose dimension is larger and possibly infinitely so. Roughly speaking, a good projection is one such that trajectories do not cross one another.[3] Takens's theorem, together with a similar one by Mañé (1981) and a more recent result by Sauer, Yorke, Casdagli and Kostelich (1990), provides a condition on n sufficient to have such a good projection. If m is the dimension of the original attractor, the condition reads:[4]

$$n \geq 2m + 1. \tag{10.2}$$

Notice that (10.2) is a sufficient condition and that in practice it is quite possible to obtain properly reconstructed attractors with embedding dimensions much smaller than n. It is known, for example, that in the reconstruction of the Lorenz attractor (whose capacity dimension is slightly larger than 2) from pseudo-experimental data (i.e., from a single time series derived from the solution of the model), $n = 3$ is quite sufficient to reproduce the essential features of the system.

10.2 THE RECONSTRUCTION OF THE ATTRACTOR – SOME DIFFICULTIES

The first difficulty which arises in the application of this method is that, since the dimension of the 'true' attractor is unknown, we would be naturally led to choosing a rather large value of the embedding dimension. This criterion, however, becomes rapidly unworkable in higher dimensions. A further problem derives from the fact that the reconstruction produces an artificial stretching of the attractor, which gets squeezed onto the bisectrix of the projection space, all the more so if the sampling time is short. On the other hand, as we shall see later, a short sampling time seems to be a necessary condition for effectively reconstructing the attractor. This is a serious problem, especially in economics where the available series are sampled at given intervals and

---

[3] In common parlance, this can be rephrased by saying that, although the number of variables of the real system whose dynamics correspond to a certain time series of GNP may be indefinitely large, a smaller number of 'essential' variables can be found which control the evolution of the system. Notice, however, that the essential variables need not correspond to any of the original physical variables, but are usually combinations of them.

[4] The inequality above applies when by 'dimension of A' we mean the 'capacity' dimension. If instead the Hausdorff dimension ($\dim_H$) is used, the inequality must be changed as follows

$$\dim(E) \geq \dim_H(A \times A) + 1.$$

where the sampling time cannot be optimally chosen. This means that it may be difficult to have a correct idea of the shape of the 'true' attractor from the observation of its projections on different subspaces of the embedding space.

However, in the attempt to verify whether a real time series has been generated by an unknown, deterministic dynamical system, the most serious problem is the inevitable presence of noise in the data. As the analysis in the present case is almost entirely based on numerical calculations, this difficulty, if not properly handled, could prove fatal. Noise makes itself felt in a particularly evident manner in economics where time series are never precise, especially when aggregates (e.g., GNP, global consumption, global investment, etc.) are concerned. Moreover, in economics, contrary to what happens in other disciplines, experiments cannot be duplicated, and therefore only one measurement of the relevant variables may be utilized in each particular instance.

The analysis of real time series from the point of view of chaos theory mainly consists in the calculation of some indicators concerning the hypothetical 'reconstructed' attractors, e.g., dominant Lyapunov exponent, fractal dimension, or spectral density. Whenever the noise present in the data goes beyond a certain threshold, and there are good reasons to believe that in economics this is the rule rather than the exception, the calculation of those indicators is affected in a significant way. Thus when estimating the dominant Lyapunov exponent from noisy data, we generally obtain estimates which are both non-converging and artificially large. The estimate of correlation dimension will also be non-convergent and will tend to increase together with the dimension of the embedding, as would be the case if the data were perfectly random. Finally, the estimate of spectral density will tend to be similar to that of a random process; that is to say, when the series is statistically stationary, it will be more or less constant except for certain estimation errors.

From these considerations, it follows that the exercise in question may become rather pointless unless we first manage to eliminate the noise which contaminates the data. In the next section, we shall discuss and make use of a method called *singular spectrum analysis* (henceforth SSA), which has proved very useful in overcoming some of the difficulties just mentioned and in extracting qualitative and quantitative information on the (assumed) underlying deterministic structure of the data, especially when these are strongly contaminated by noise.

10.3 SINGULAR SPECTRUM ANALYSIS

The SSA method is based on a particular transformation known in matrix algebra as 'singular value decomposition' and presents strong similarities with the analysis of the principal components of a linear model. However, its application does not presuppose any *a priori* hypothesis on the form of the system which is assumed to have generated the data.

To outline the essential features of the method, let us go back to (10.1) and Theorem 10.1. The space which contains the image of the map $\Phi_{F,v}$ is called the embedding space and its dimension, denoted by n, is called the embedding dimension. (Theorem 10.1, of course, requires that $n \geq 2m+1$ where m is the dimension of the attractor.) We can think of (10.1) as a sampled form of a function $Y(t)$, consisting of measurements $Y_k = Y(k\tau_s)$, sampled regularly in time for $k = 1, \cdots, n$, where τ_s is the sampling interval. We can now introduce the concept of an '(n, τ_s)-window' which makes visible n elements of the time series, sampled at intervals of length τ_s. If the original time series is discrete, then the smallest value of τ_s is 1. In this case the window shows n consecutive elements and we speak simply of an 'n-window'. If $Y(t)$ is continuous, τ_s can be made arbitrarily small and the number of measurements which can be derived from any finite time interval is arbitrarily large. The quantity $\tau_w = n\tau_s$ is called the 'window length'. In principle, the embedding dimension n can be increased either by increasing the number of elements for any given sampling interval (i.e., by increasing τ_ω), or by reducing the sampling interval itself, given the window length. In practice, however, available data are limited both in the sense that the time interval is bounded and that the choice of sampling interval is drastically restricted. This of course puts severe limitations on the time scales ('extensive' and 'intensive') of the dynamics which can be reconstructed. Once an (n, τ_s)-window is defined, as the time series is advanced through it stepwise, a sequence of vectors in the n-dimensional embedding space is generated, namely, $\{x_i \in \mathbb{R}^n | i = 1, \cdots, N\}$, where $N = \tilde{N} - n + 1$, $\tilde{N}$ being the number of univariate observations. Those sequences, which are called *n-histories*, can be used to construct an $N \times n$ matrix which will henceforth be called the *trajectory matrix*, i.e.,

$$X = N^{-1/2} \begin{pmatrix} x_1^T \\ \vdots \\ x_N^T \end{pmatrix}, \tag{10.3}$$

where the factor $N^{-1/2}$ has been introduced for convenience's sake. By plotting the columns of X against the principal directions of the embedding space (and assuming that $n \geq 2m + 1$), we obtain the reconstructed attractor.

The SSA has been well known for some time in signal theory and its application to dynamical system theory has recently been suggested by Broomhead and King (BK) (1986a,b) from whose analysis we shall start.

There exists, however, a better way of projecting the reconstructed attractor which, as we have already said, can overcome some of the difficulties of the Takens method. For this purpose, we consider a set of vectors $\{s_i \in \mathbb{R}^N | i = 1, \cdots, n\}$ such that by their action on X they generate a set of linearly independent and orthogonal vectors $\{c_i \in \mathbb{R}^n | i = 1, \cdots, n\}$.

Without any loss of generality, we can assume that the vectors $\{c_i\}$ are also normal and consequently give rise to an orthonormal basis for $\mathbb{R}^n$. From this, we can derive the following relation:

$$s_i^T X = \sigma_i c_i^T \tag{10.4}$$

where $\{\sigma_i\}$ is the set of real constants employed to normalize the vectors. From linear algebra we know that orthonormality of $\{c_i\}$ requires that

$$s_i^T X X^T s_j = \sigma_i \sigma_j \delta_{ij}, \tag{10.5}$$

where δ is the Kronecker delta. The matrix $H = X X^T$ is real and symmetrical and therefore its eigenvectors form an orthonormal basis for $\mathbb{R}^N$. It is precisely the eigenvectors of H that satisfy (10.5), hence we have:

$$H s_i = \sigma_i^2 s_i. \tag{10.6}$$

We have thus determined the set $\{s_i\}$ (the eigenvectors of H) and also the set $\{\sigma_i^2\}$ (the corresponding eigenvalues, all real and non-negative since H is positive-semidefinite). The form of H is the following:

$$H = N^{-1} \begin{pmatrix} x_1^T x_1 & \cdots & x_1^T x_N \\ \vdots & \vdots & \vdots \\ x_N^T x_1 & \cdots & x_N^T x_N \end{pmatrix}, \tag{10.7}$$

and therefore it can be interpreted as the matrix which describes the correlation between all the pairs of vectors generated by a window of length n, used to construct the n-histories.

We still have to face the rather formidable problem that H has dimension $N \times N$, i.e., in general, it is rather large and consequently its diagonalization may be practically impossible.

A more efficient way of obtaining the desired result is the following. Taking the transpose of (10.4) we obtain:

$$X^T s_i = \sigma_i c_i \tag{10.8}$$

from which, premultiplying by X, we have:

$$XX^T s_i = \sigma_i X c_i. \tag{10.9}$$

Finally, by making use of (10.6) and simplifying, we have:

$$X c_i = \sigma_i s_i. \tag{10.10}$$

If we now premultiply (10.10) by X^T and take into account (10.4), we obtain the equation

$$V c_i = \sigma_i^2 c_i, \tag{10.11}$$

where $V \equiv X^T X \in \mathbb{R}^{n \times n}$ is a real symmetric, positive-definite matrix. Equation (10.11) permits us to derive the vectors $\{c_i\}$ as eigenvectors of V, and $\{\sigma_i^2\}$ as the corresponding eigenvalues. Notice that, putting the mean equal to zero, V represents the matrix of covariances of the observations which are included in the vectors x_i, calculated over the entire trajectory. The number of lags will be equal to the dimension of the embedding, and, if the sampling interval is one, to the length of the window. Notice that equation (10.11) is much easier to deal with than (10.6), since n is usually much smaller than N.

If we now consider the matrix C whose columns are formed by vectors c_i, and $\Sigma^2 \equiv \mathrm{diag}(\sigma_1^2, \ldots, \sigma_n^2)$ where σ_i^2 are ordered from the largest to the smallest, we can see that (10.11) can be written as

$$VC = C\Sigma^2. \tag{10.12}$$

Using the definition of V, we have

$$(XC)^T (XC) = \Sigma^2. \tag{10.13}$$

The matrix XC represents the trajectory matrix projected onto the basis $\{c_i\}$. If we think of a trajectory as exploring on average an ellipsoid of dimension n, the vectors c_i $(i = 1, \cdots, n)$ correspond to the directions of the principal axes of the ellipsoid, and the values σ_i^2 associated with them correspond to the lengths of those axes.

The choice of $\{c_i\}$ as a basis for the projection (i.e., to project the trajectory matrix onto the space spanned by the eigenvectors of the covariance matrix of the time series), is optimal in the sense that it makes the columns of the trajectory matrix independent (see (10.13)) and minimizes the mean square error of projection. In this way, the plots are no longer squeezed onto the diagonal and the projections on planes (i, j) and $(i + k, j + k)$ are no longer equal as would be the case with the Takens method. It should also be noticed that this method introduces a first, rough, filtering of the series, as the projection of the matrix X onto the basis $\{c_i\}$ can be viewed as a weighted moving average of the original data.

10.4 FILTERING AWAY NOISE

We have so far proceeded as if the series under investigation were 'clean' of noise or, more precisely, as if the only noise derived from finite sampling. However, the crucial property of the SSA method is its ability to filter away (some of) the noise present in the data thereby contributing to the identification of the directions along which the deterministic component of motion takes place, which we shall henceforth call 'significant (or deterministic) directions', whereas the infinite rest will be denoted by 'stochastic directions'. In order to explore this problem, we shall assume that we are dealing with a time series characterized by zero mean, continuity[5] and ergodicity.

When looking for low-dimensional deterministic chaos in a time series, we try to show that the series has been generated by a dynamical system whose orbits are confined to a subset of the phase space, whose dimension is, say, d. However, if the data are contaminated by noise, the orbits will in fact move along infinite directions, i.e., noise can be viewed as an infinite-dimensional process. We could hope to be able to divide the embedding space, i.e., the space in which the attractor is immersed, into a deterministic subspace where the orbits would stay in the absence of noise, and a stochastic subspace, in which the motion is due only to noise. In their work, BK claim that the SSA method provides significant information about the dimension d of the deterministic subspace, and they reduce the search for the value of d to an eigenvalue problem. Indeed – so the argument runs – in the noiseless case, d is nothing but the number of linearly independent vectors which can be constructed from the trajectory matrix. Therefore, in order to determine d, it is sufficient to evaluate the rank of the matrix H (or V). That is to say, d will be equal to the number of positive eigenvalues of V, the other $(n - d)$ eigenvalues being equal to zero.

The presence of noise (at least in the simple case in which it is white) would simply lift all the positive eigenvalues by a given amount which corresponds to the 'noise floor'. In fact, if we arrange the eigenvalues of a matrix H (or V) constructed from experimental data in descending order, we shall observe a finite, possibly small, number of distinct 'emerging' eigenvalues, plus an (approximately) flat tail.[6]

---

[5] That is to say, we treat the discrete data available as samples taken at a regular interval from a continuous time series.

[6] Of course, if the noise is strong, it will overwhelm the signal in the directions associated with small eigenvalues and the signal will not be distinguishable from the floor.

According to BK, the rank of V 'is an upper bound to the dimensionality of the subspace explored by the deterministic component of the trajectory' (BK, 1986a, p. 225). Thus, BK interpret the d 'emerging' eigenvalues of H (or V) as the dimension of the deterministic subspace.

Unfortunately, it has been shown[7] that this reasoning is too simplistic. Firstly, if the data can be viewed as samples of a continuous signal, the matrix H (or V) has generically full rank, although the eigenvalues tend to zero as n tends to infinity. The rank of V, therefore, goes to infinity with the dimension n of the embedding space, even in the absence of noise. Moreover, and more importantly, the level of the noise floor is not independent of the embedding dimension n. When we increase the embedding dimension, either by increasing the window length or by decreasing the sampling interval, the noise floor is lowered and new 'emerging' eigenvalues appear.

To understand this point better, let us rewrite equation (10.11) as

$$\theta V c_i = \sigma_i^2 c_i, \tag{10.14}$$

where $\theta = 1/n$ is a scaling factor chosen for convergence purposes, as we shall see in a moment. Then, in the presence of noise, the estimated eigenvalues of (10.14) are

$$\sigma_i^2 = (\sigma_i^T)^2 + \frac{1}{n}\sigma_w^2,$$

where $(\sigma_i^T)^2$ denotes an estimate for the signal corresponding to the 'true', clean series, and σ_w^2 measures the variance due to noise. If we had infinitely precise information about the system, for $n \to \infty$, σ_i^2 would converge to the 'true' signal.[8]

The noise floor is given by σ_w^2/n and clearly depends on n. The latter, however, can be expressed as a function of the window length and the sampling rate thus:

$$n = \frac{\tau_w}{\tau_s}.$$

Therefore, for a fixed τ_w, n depends inversely on τ_s.

The unfortunate consequence of all this is that we cannot accurately estimate the significant deterministic directions of the system in any simple manner, either by identifying the rank of the covariance matrix V (in the noiseless case), or by identifying the eigenvalues situated above the noise level.

---

[7] See Vautard and Ghil (1989), on which the following paragraphs are based.

[8] Also if $n \to \infty$, while τ_w is kept constant, the sampled signal would converge to the continuous one.

Therefore, the number of eigenvalues above the noise floor, which Vautard and Ghil call 'statistical dimension' and is denoted here by SD, depends on the quality and quantity of data, and from it we cannot immediately derive an estimate of the true dynamical dimension, i.e., the dimension of the attractor. The statistical dimension SD, however, is a very useful concept, for the following reasons.

(1) It provides a reliable upper bound for the minimum number of degrees of freedom necessary to reconstruct the time series under investigation. That is to say, SD is the minimum number of distinct variables and functional relationships between them, necessary to formulate a model whose output can then be compared with those data. From this point of view, it is little wonder that SD must be a function of the accuracy of the data.

(2) It helps us verify the estimates of the fractal dimension D of the (hypothetical) attractor of the system. Suppose that the estimate of D saturates at a value $\bar{D}_c$ corresponding to an embedding dimension $\bar{n}$.[9] Then, in order for $\bar{D}_c$ to be a correct estimate of D, it is necessary (but not sufficient) that

(i) $\bar{D}_c \leq SD(n) \leq n$;

(ii) $2\bar{D}_c + 1 \leq n \leq \bar{n}$.

(3) So far we have only considered the role of the 'emerging' *eigenvalues* of the covariance matrix V, but further information is carried by the corresponding *eigenvectors*. The latter can be viewed as moving average filters determined directly from the data. They can in fact be used to filter away (part of) the noise in the data, in order to more accurately reconstruct the attractor and calculate its statistical properties. Broomhead, Jones and King (1987) also suggest the inspection of the eigenvectors of the matrix V in order to identify the deterministic and the stochastic directions of motion. According to these authors, the shape of the 'deterministic eigenvectors'[10] should be regular and reminiscent of that of orthogonal polynomials of degree $i - 1$, i being the order of the eigenvalue corresponding to the eigenvector in question. On the contrary, 'stochastic eigenvectors' should have a noisy shape. Although a rigorous

---

[9] Cf. our discussion in Chapter 7.

[10] By 'shape of an eigenvector' consisting of n elements, we here mean a curve in a plane, constructed by interpolating the points whose two co-ordinates are given by the value of an element of the vector and the number corresponding to its position in the vector.

mathematical explanation of this fact does not seem to us to have been provided, the idea works surprisingly well in practice, as we shall see later.

We shall conclude by giving a concise description of the actual calculations required to apply the filtering method discussed so far, according the procedure indicated by BK and followed by us in the exercises of Chapter 14.

From linear algebra, we know that a rectangular matrix (in this case the trajectory matrix X) can be decomposed by means of the so-called singular value decomposition in the following way:

$$X = S\Sigma C^T, \; S \in \mathbb{R}^{N \times n}, \; \Sigma \in \mathbb{R}^{n \times n}, \; C^T \in \mathbb{R}^{n \times n}. \qquad (10.15)$$

Assuming that the d significant directions have been more or less correctly identified, we can now split the matrix Σ as follows:

$$\Sigma = \Sigma^* + \hat{\Sigma}, \qquad (10.16)$$

where Σ^* is obtained from Σ by putting the $n - d$ eigenvalues belonging to the noise floor equal to zero, and $\hat{\Sigma}$ is also obtained from Σ by putting the d significant eigenvalues equal to zero. Specifically, we shall write:

$$\Sigma^* = \text{diag}(\sigma_1, \ldots, \sigma_d, 0, \ldots, 0) \qquad (10.17)$$

$$\hat{\Sigma} = \text{diag}(0, \ldots, 0, \sigma_{d+1}, \ldots, \sigma_n). \qquad (10.18)$$

If we replace the matrix Σ in (10.15) by Σ^* or $\hat{\Sigma}$, we obtain, respectively, the deterministic and the stochastic components of the trajectory matrix, namely:

$$X^* = S\Sigma^* C^T, \qquad (10.19)$$

$$\hat{X} = S\hat{\Sigma}C^T. \qquad (10.20)$$

Adding the equations (10.19) and (10.20), we can verify that

$$X = X^* + \hat{X}, \qquad (10.21)$$

i.e., the sum of the two separate components is indeed equal to the original trajectory matrix. As we know (subject to the qualifications discussed in the previous sections), from a geometrical point of view, these operations can be interpreted by saying that X^* represents that part of the trajectory which moves in the deterministic subspace. In other words, from the original matrix we have filtered away the motion along the $n - d$ non-deterministic directions, which is due to noise and is represented by the matrix $\hat{X}$.

In spite of its apparent simplicity, equation (10.19) is hardly operational because of the large dimension of S. Substituting (10.10) into (10.19), we shall obtain the much simpler and more practical expression:

$$X^* = \sum_{i=1}^{d} [Xc_i c_i^T],$$
(10.22)

where X^* is of course a $(N \times n)$ matrix. In constructing equation (10.22), we make use only of the original data and of quantities derived from the diagonalization of V. Since we are actually interested in the d deterministic directions, we can project the trajectory matrix onto the deterministic subspace spanned by the eigenvectors corresponding to the eigenvalues above the noise floor, by performing the following operation:

$$\bar{X} = X^*C.$$
(10.23)

In this manner, as we have already seen, we shall also obtain an optimal projection of trajectories.

From now on, all the calculations necessary to decide whether the time series under scrutiny has been generated by a chaotic system will clearly have to be performed on $\bar{X}$. We would like to point out that overestimation of the number of significant eigenvalues may not be too dangerous a mistake. Indeed, allowing for one or two excess directions is equivalent to leaving some noise in the data, but this will certainly be less than the amount present before the filtering. We will have something more to say about this residual noise when, in Chapter 14, we try to estimate certain indicators of chaos, such as Lyapunov exponents or correlation dimensions, in relation to an economic time series.

Part II

Applications to economics

Part II

11

Discrete and continuous chaos

11.1 INTRODUCTION

The majority of applications of chaotic dynamics to economic problems[1] consist of adaptations, with various amounts of value-added, of the celebrated first-order, nonlinear difference equation, so brilliantly studied by Robert May (1976) and subsequently investigated in a large number of papers and books.[2]

As is well known, the May equation has the general form

$$x_{t+T} = G(x_t), \tag{11.1}$$

where $x \in \mathbb{R}$, $G: \mathbb{R} \to \mathbb{R}$ is a smooth 'one-hump' function, and T is the length of a fixed delay, which of course can always be made equal to one by appropriately choosing the unit of measure of time. We have already seen that the behaviour of the discrete dynamical system described by (11.1), quite independently of the specific form of $G(\cdot)$, essentially depends on a single parameter and, over a certain interval of values of this parameter, may display chaotic behaviour. The question which immediately leaps to mind is the following: how general are the (essentially mathematical) results derived from investigation of the discrete dynamical system (11.1), and how relevant are they to economic theory?

Notice, first of all, that a system described by equation (11.1) may be conceived as a highly aggregate mechanism (a one-loop feedback system) consisting of two parts: a nonlinear functional relationship and a lag, the latter being, in this particular case, a fixed delay. (See Fig. 11.1.)

Let us consider these two elements in turn. The nonlinearities implied by single-hump functions are of a rather common kind and their presence has been detected in situations pertaining to practically all branches

---

[1] Some of the relatively few exceptions are discussed in Lorenz (1989b).
[2] Cf. our discussion of one-dimensional maps in Chapter 9, and the works cited there.

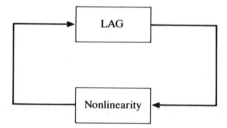

Fig. 11.1. A one-loop feedback system.

of the natural and social sciences. Economists may (and indeed do) disagree on the likelihood of such nonlinearities actually affecting the operation of real economic systems, and some of them are busy proving or disproving their existence in specific cases.[3] There should be consensus, we believe, that the general laws governing rational economic behaviour, as postulated by the prevailing theory, do not exclude *a priori*, and indeed in some cases logically imply, the presence of non-monotonic functional relationships of the one-hump type.

This point can be appreciated by considering some of the applications to economics of equations of the type (11.1), including, amongst others macroeconomic models (e.g., Stutzer, 1980; Day, 1982); models of rational consumption (e.g., Benhabib and Day, 1981); models of overlapping generations (e.g., Benhabib and Day, 1982; Grandmont, 1985); models of optimal growth (e.g., Deneckere and Pelikan, 1986). Recent overviews of the matter with further instances of one-hump functions derived from economic problems can be found in Baumol and Benhabib (1988), Lorenz (1989b), Boldrin and Woodford (1990) and Scheinkman (1990).

On the contrary, the role of the second element of the May model, i.e., the lag, has been rather neglected in economists' discussions of chaotic dynamics. Comments on this point are often limited to some scattered observations that most mathematical theorems utilized in this case only apply to discrete, one-dimensional systems, and that the most interesting result (the occurrence of chaotic dynamics) disappears when (supposedly) equivalent continuous-time formulations of the same problems are considered.

This neglect is particularly surprising since there exists in the economic literature a lively and intellectually stimulating debate on the relative

---

[3] Cf. our discussions in Chapters 10 and 14, and the bibliography given there.

virtues and shortcomings of discrete and continuous models, which is very relevant to the point in question. Indeed, this chapter may be considered an attempt to clarify some obscure aspects of this unresolved issue.

For this purpose, it will be expedient to briefly recall the main difficulties that arise when one studies an aggregate model (economic or otherwise), by means of so-called 'period analysis', i.e., in discrete time.[4] First of all, even though economic transactions of a given type do not take place continuously and are therefore discrete, in general they will not be perfectly synchronized as period analysis implicitly assumes, but will overlap in time in some stochastic manner. Only in very rare circumstances (e.g., in an agricultural, single-crop economy) could one define a 'natural period' for the economic activity under investigation. Whenever this is not possible, there is a danger that the implicit assumption underlying the fixed-delay hypothesis may yield misleading conclusions.

The danger that some of the conclusions obtained by means of period analysis may be mere artifacts owing to a misspecification of the model is clearly present in virtually all the existing applications of equation (11.1) to economic problems for which no aggregate 'natural period' could be defined. Those who think that chaotic dynamics do exist in real economies (and these authors are among them) are therefore under obligation to show how this particular misspecification can be avoided.

In what follows, therefore, we shall discuss a procedure that, in our opinion, sheds light on the connection between discrete- and continuous-time representations of a process. In so doing, moreover, we shall apply some of the theoretical concepts discussed in Chapter 2, as well as some of the numerical methods illustrated in Chapters 4–9 to a specific economic problem.[5]

11.2 PROBABILITY DISTRIBUTIONS OF LAGS

Let us consider any of the economic models discussed in the works cited in the preceding section (perhaps Benhabib and Day's model of rational consumption, or Grandmont's overlapping-generations model),[6]

---

[4] This paragraph closely follows the discussion developed in J. May (1970) (a different May!), Foley (1975) and Turnovsky (1977). In order to facilitate reference to economic literature, in what follows we use the concepts of 'period' and 'fixed-delay lag' as synonyms.

[5] In the next few paragraphs, we follow Invernizzi and Medio (1991), to which we refer the reader for a more detailed discussion and bibliography.

[6] As concerns this particular model, however, we only refer here to the so-called 'backward

and suppose that, although we accept the authors' arguments in every other respect, for the reasons discussed above we reject the hypothesis of a single fixed delay as an unduly crude way of aggregating the economy.

Instead of a single 'representative' economic agent (or unit), as implicitly postulated by those models, we consider a hypothetical economy consisting of an indefinitely large number of agents, who respond to a certain signal with given discrete lags. The lengths of the lags are different for different agents, and are distributed in a random manner over all the population. In this situation, the economy's aggregate time of reaction to the signal may be modelled by a real nonnegative random variable T, the overall length of the delay. When observations on T can be repeated indefinitely, the frequency ratio for any specific range of values will tend to the corresponding probability. The set of observations approached in the limit is the *population* of our observations. In this respect, therefore, every hypothetical economy may be characterized by the continuous, essentially positive, nonconstant random variable T, whose distribution can be estimated through the usual statistical procedures.

A continuous random variable is known if its probability distribution is known. In principle, the overall reaction time T can be distributed in a number of different ways, depending on the specific problem at hand.

There exist, however, certain general criteria for choosing an 'optimal' probability distribution, given certain constraints. In the attempt to estimate the true distribution of a random variable, statisticians, in order to avoid unjustified biases, should formulate their assumptions so as to maximize the uncertainty about the system, subject to the constraints deriving from prior knowledge of the problem. A rigorously defined measure of uncertainty (i.e., lack of information) is provided by 'entropy'. The concept of entropy and its relationship with those of information and predictability has already been mentioned in Chapter 2. The 'principle of maximum entropy' for selecting a probability distribution was put forward in the economic literature by Theil and Fiebig (1981) and can be considered as a generalization of Laplace's famous 'principle of insufficient reason'.

In the case of essentially positive random variables like lags, for example, if the only constraint on the probability distribution is that the range of T is a bounded interval, then we should choose the uniform distribution.

dynamics', leaving aside any questions concerning its relation with the 'true', forward dynamics. This point, however, will be taken up again in Chapter 12.

If, on the other hand, the range of T is $(0, \infty)$ and the only constraint is that the expected value, or mean, of T exists and is equal to τ, then the principle of maximum entropy requires that we choose the (one-parameter) exponential distribution, with parameter $\theta = 1/\tau$. Moreover, if an additional independent constraint is assumed, and if we also fix the geometric mean, the maximum entropy criterion requires the choice of a two-parameter gamma distribution.

We can recall that, if we indicate the shape parameter by α and the scale parameter by β, the density of a two-parameter gamma random variable can in general be written thus:

$$g(\alpha, \beta; t) = \begin{cases} 0, & \text{if } t \leq 0, \\ \{\beta^{\alpha}\Gamma(\alpha)\}^{-1}t^{\alpha-1}e^{-t/\beta}, & \text{if } t > 0, \end{cases} \tag{11.2}$$

where Γ is the gamma function, i.e.,

$$\Gamma(\alpha) = \int_0^{\infty} e^{-t}t^{\alpha-1}dt$$

and $\alpha, \beta > 0$. From the knowledge of the parameters α and β, we can derive the mean $(\alpha\beta)$, the variance $(\beta^2\alpha)$ and the geometric mean $(\beta \exp(\Gamma'(\alpha)/\Gamma(\alpha)))$ of the distribution.

Conversely, it can be shown that, for the gamma distribution, fixing mean and geometric mean is equivalent to fixing mean and variance. Since the latter two statistical indicators are the most commonly used in economics, we shall henceforth use them as the parameters of the distribution.

11.3 A FEEDBACK REPRESENTATION OF LAGS

Let us now abandon for a moment the probabilistic aspect of lags, and consider the question from a feedback point of view.

Suppose a variable $Y(t)$ is related with a continuously distributed lag to another variable $Z(t)$, where of course $Z(t)$ may well indicate the same variable Y at some time different from t. The equation of the lag can be written in its time form thus:

$$Y(t) = \int_0^{\infty} w(s)Z(t-s)ds, \tag{11.3}$$

where $w: \mathbb{R} \to \mathbb{R}$ is continuous and we have

$$w(\sigma) = 0, \text{ for } \sigma \leq 0; \ w(\sigma) \geq 0, \text{ for } \sigma > 0; \text{ and } \int_0^{\infty} w(s)ds = 1.$$

Thus w can be thought of as a 'weighting function': it indicates the strength of impact that values of Z in the more or less distant past have on the value of Y now. In principle any kind of time profile of such an impact can be so modelled.

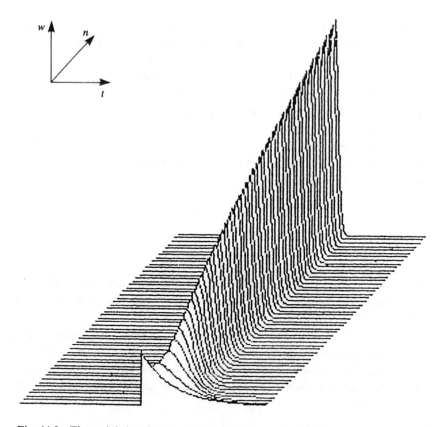

Fig. 11.2. The weighting function of a multiple exponential lag as a function of
the parameter n.

Under rather general circumstances, the lag can be represented equivalently in the form of a differential operator. If we restrict the variable
Z to be zero for $t < 0$, then, for each fixed t, $w(s)Z(t - s)$ is also zero for
all $s > t$. Therefore (11.3) can be rewritten thus:

$$Y(t) = \int_0^t w(s)Z(t - s)ds. \tag{11.4}$$

The expression on the RHS of (11.4) arises in the theory of the Laplace
transform as the *convolution* of w and Z, and it is denoted by $w * Z$. We
recall that $w * Z = Z * w$. Moreover, under general conditions (see, for
example, Kaplan, 1962, p. 340), we have:

$$\mathscr{L}[w * Z] = \mathscr{L}[w] \cdot \mathscr{L}[Z], \tag{11.5}$$

where $\mathscr{L}$ is the usual Laplace transform.

If, moreover, $\mathscr{L}[w]$ is the reciprocal of a polynomial

$$V(p) = a_0 p^n + \cdots + a_n,$$

that is

$$w(s) = \mathscr{L}^{-1}[1/V(p)],$$

then $Y = w * Z$ is exactly the unique solution of the differential equation

$$V(D)Y(t) = Z(t), \qquad (11.6)$$

where $D \equiv \mathrm{d}/\mathrm{d}t$, with initial conditions $Y = Y' = \cdots = Y^{n-1} = 0$ at $t = 0$. (Cf. Kaplan, 1962, p. 358, th. 20.)

In particular, if the lag is a multiple exponential lag of order n, we have:

$$\frac{1}{V(p)} = \left(\frac{\tau p}{n} + 1\right)^{-n}$$

and consequently (11.6) will take the form:

$$\left(\frac{\tau D}{n} + 1\right)^n Y(t) = Z(t),$$

or

$$Y(t) = \left(\frac{\tau D}{n} + 1\right)^{-n} Z(t), \qquad (11.7)$$

where n is a positive integer and τ is the time constant of the lag.

If we now calculate the 'weighting function' $w(t)$ corresponding to a multiple exponential lag by performing an inverse Laplace transform operation,[7] we obtain

$$w(t) = \left(\frac{n}{\tau}\right)^n \frac{t^{n-1}}{(n-1)!} e^{-nt/\tau} \qquad (11.8)$$

Inspection of (11.2) and (11.8) will promptly reveal that the weighting function of a multiple exponential lag of order n is the same as the density function of a two-parameter gamma distribution.[8] The time constant τ of the lag operator corresponds to the mean, and the order of the lag, n, is inversely related to the variance. It can also be seen immediately that, for $n \geq 2$, $w(t)$ will have a 'one-hump' form, with a single critical point at $t = \tau(n-1)/n$.[9] The greater the order of the lag, the smaller will be the variance around the mean τ. For large values of n, therefore, the weighting function will have a peaked form, indicating that the value of

---

[7] For details of this operation, see, for example, Doetsch (1971, ch. 3).

[8] To the best of our knowledge, this result was first established by Invernizzi and Medio (1991).

[9] The relevance of this comment will become apparent in our discussion in Chapter 13.

the output at any given instant t mostly depends on the input delivered at a given instant of the past, $t - \tau$, or in a certain small neighbourhood around it. All this can be appreciated by looking at Fig. 11.2.

Two simple cases can be most often found in the economic literature, namely:

(i) *The simple exponential lag*, corresponding to $n = 1$. In this case, the reaction to the input starts immediately at $t = 0$, but the temporal evolution of the output, that is the growth of Y, is proportional to the excess of the input over the output. This leads to the following ordinary differential equation:

$$\dot{Y}(t) = \gamma[Z(t) - Y(t)], \tag{11.9}$$

where γ (> 0) has dimension *time*$^{-1}$ and represents the *speed of adjustment*, whereas its inverse $1/\gamma = \tau$ can be considered the length of the simple exponential lag. In economic applications, the variable Z sometimes represents the desired, or equilibrium value of Y, so that (11.9) depicts an 'Achilles and the Tortoise' situation in which the actual magnitude *chases* the desired one, approaching it at an exponentially slowing speed, and catching up with it only in the limit as $t \to +\infty$.

The simple exponential lag, or exponential lag of order one, corresponds to a rather special (and crude) formalization of economic reaction mechanisms.

(ii) *The fixed delay.* When n becomes indefinitely large, the weighting (and the density) function tends to a Dirac delta function on τ, and the continuous exponential lag tends to a fixed delay of length τ. In fact, we have

$$\lim_{n \to \infty} \left(\frac{\tau D}{n} + 1 \right)^{-n} = e^{-\tau D}. \tag{11.10}$$

The expression on the RHS of (11.10) is called a *shift operator*, which, when applied to a continuous function of time, has the effect of translating the entire function forward in time by an interval equal to τ.[10] Thus the fixed-delay lag employed in models of the type (11.1) can be seen as a special, limit case of a multiple exponential lag when the order of the lag tends to infinity. Equivalently, the aggregate fixed delay of those models can be viewed as a limit case of a system characterized by gamma-distributed individual reaction times, which obtains when the dispersion around the mean (the variance) tends to zero.

---

[10] Cf. Yoshida (1984, part III, ch. 8).

The question now naturally arises whether some or all the interesting results obtained by the investigation of one-dimensional maps of the type (11.1) can be reproduced by means of models adopting less drastic and more flexible assumptions concerning the weighting function, and consequently the time profile of the lag.

Mathematically, this can be done by replacing (11.1) with the more general, continuous-time nth order differential equation[11]

$$x_n = \left(\frac{D}{n} + 1 \right)^{-n} G(x_n), \qquad (11.11)$$

or, equivalently, by the system of n first-order differential equations:

$$\left(\frac{D}{n} + 1 \right) x_j = x_{j-1}, \quad j = 2, \dots, n, \qquad (11.12)$$

$$\left(\frac{D}{n} + 1 \right) x_1 = G(x_n), \qquad (11.13)$$

where $G(\cdot)$ is a one-humped function like (11.1), $D \equiv d/dt$ and, for simplicity's sake, τ has been put equal to one. The main purpose of what follows is to show that some of the most interesting features of (11.1) can be reproduced by the system (11.12) and (11.13) at a reasonably low level of dimensionality (i.e., for relatively low values of n). In particular, we wish to show that chaotic behaviour, which is known to exist for the former, also exists for (relatively) low-dimensional specifications of the latter. To made things more specific, from now on we select a particular form of the function $G(x_n)$ which appears in (11.13), namely:

$$G(x_n) = rx_n(1 - x_n).$$

Qualitatively similar results can be obtained with different specifications of the one-hump function (see Invernizzi and Medio, 1991).

11.4 SYMPTOMATOLOGY OF CHAOS

Equipped with the ideas and tools of analysis investigated in the previous chapters, we first discuss some properties of the system (11.12) and (11.13) which are independent of n.

From (11.12) and (11.13), we gather that the equilibrium conditions are the following:

$$x_1 = x_2 = \cdots = x_n = \bar{x},$$

---

[11] A similar approach to this problem can be found in a paper by Sparrow (1980), from the study of which we have greatly benefited. In fact, some of the results in this section and the following one can be considered as further developments of the same line of research. See also May's comments on this point (1983, pp. 548–9, 555).

and

$$\bar{x} = r\bar{x}(1 - \bar{x}),$$

whence we obtain the two equilibrium solutions:

$$\bar{x}_1 = 0; \quad \bar{x}_2 = 1 - (1/r). \tag{11.14}$$

As concerns the stability of equilibria, consider that the auxiliary equation can be written in the simple form:

$$(1 + \lambda)^n = G'(\bar{x}), \tag{11.15}$$

where $n\lambda$ indicates an eigenvalue of the Jacobian matrix and $G'(\bar{x}) = dG(\bar{x})/dx = r(1 - 2\bar{x})$.

Hence we have:

$$G'(\bar{x}_1) = r; \quad G'(\bar{x}_2) = 2 - r. \tag{11.16}$$

From (11.14)–(11.16) we gather that, for $r < 1$, the origin is the only nonnegative equilibrium point, and it is stable. At $r = 1$, we have a transcritical bifurcation: the equilibrium point at the origin loses its stability and a second, initially stable, equilibrium point bifurcates from it in the positive orthant of the phase space.

For $n < 3$, nothing much happens when we increase the parameter r: the positive equilibrium point remains stable for all values of $r > 1$ (for $n = 2$, damped oscillations occur for $r > 2$). For $n \geq 3$, however, as r increases past a certain value which depends on n, a Hopf bifurcation takes place and a periodic orbit bifurcates from the equilibrium point. Successive bifurcations can be detected, for greater values of r, although their exact structure still escapes us. Whatever value the parameter r may take, however, nothing more complicated than periodic orbits seems to occur for low values of n. However, when the order of the exponential lag becomes sufficiently large, there does appear to exist a value of r beyond which the system gives chaotic output.

In order to analyse this case in detail, we select $n = 10$. In this case, the first Hopf bifurcation occurs for $r \approx 3.6$ and our numerical study indicates that, when the control parameter r is increased further, the system undergoes a period-doubling sequence of bifurcations, eventually exhibiting chaotic behaviour. This can be appreciated by looking at Figs. 11.3–11.7. Fig. 11.3 shows plots of post-transient trajectories of the system for different values of r. The reader will observe the progression of complexity from a limit cycle, to period 2, 4, 8 cycles, and finally, for $r \approx 5$, to a chaotic attractor. This sequence can also be detected by means of spectral analysis of the same trajectories, as illustrated by Fig. 11.4. Starting from Fig. 11.4a we observe a basic frequency, f_0

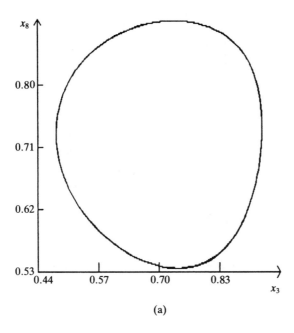

(a)

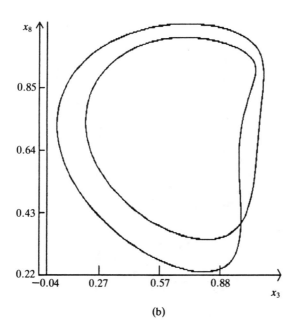

(b)

Fig. 11.3. Period-doubling route to chaos.

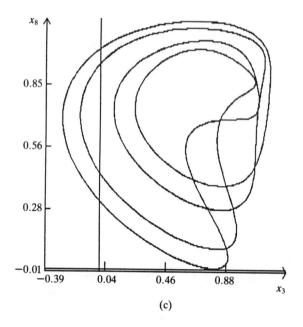

(c)

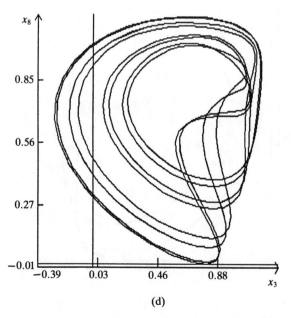

(d)

Fig. 11.3. (cont.)

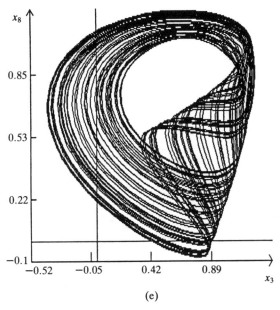

(e)

Fig. 11.3. (*cont.*)

(plus a number of harmonics). Then, increasing the parameter, we have the appearance in the spectrum of frequencies $f_0/2$, $f_0/4$, $f_0/8$. For even higher values of r, a 'noise floor' associated with chaotic behaviour appears and successively destroys the subharmonics $f_0/2^n$ in the reverse order of their appearance. In the last diagram, Fig. 11.4e, corresponding to the chaotic attractor of Fig. 11.3e, no distinct peak is left in the spectrum, which looks like a broad band.

Another interesting representation of the period-doubling route to chaos can be observed in the bifurcation diagram of Fig. 11.5.[12] The familiar period-doubling scenario is quite evident and so is the presence of periodic windows within the chaotic zone, which makes this diagram very similar to that obtained for the one-dimensional maps discussed in Chapter 9.

Finally, numerical evidence of (and quantitative information about) chaotic behaviour of the system under investigation has been provided by the estimate of the LCEs and the fractal dimension, calculated for

---

[12] To construct this diagram we have plotted, for each value of the parameter r, the maxima of the relevant variable, after discarding some transients.

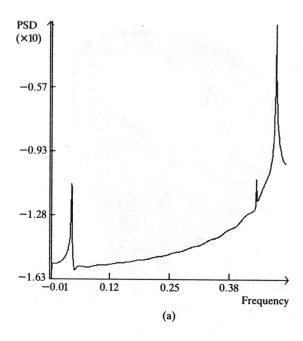

(a)

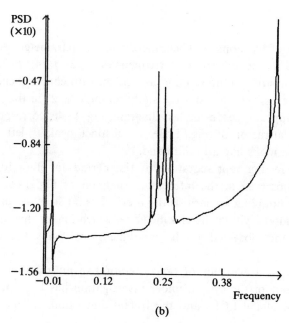

(b)

Fig. 11.4. PSD of the period-doubling sequence.

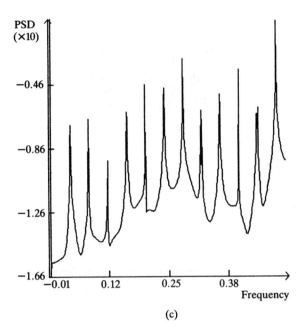

(c)

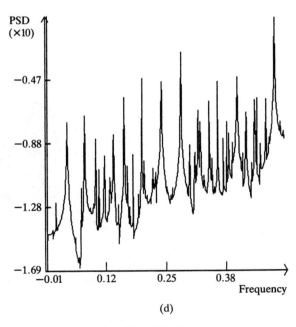

(d)

Fig. 11.4. (*cont.*)

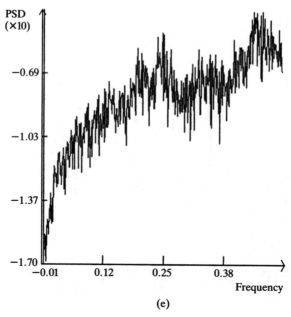

(e)

Fig. 11.4. (*cont.*)

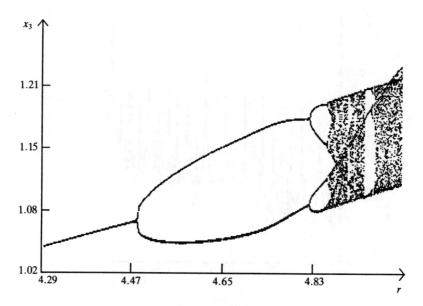

Fig. 11.5. Bifurcation diagram.

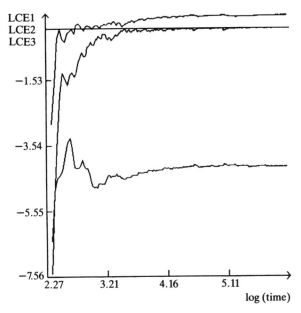

Fig. 11.6. LCEs in the chaotic zone.

$r = 5$, i.e., within the chaotic zone. As we know from the analysis of Chapter 6, the existence of an attractor with one or more positive LCE indicates that the motion of the system on the attractor has sensitive dependence on initial conditions, i.e., it is chaotic. Since in our case the attractor looks approximately two-dimensional, we would expect the sign pattern of the ten LCEs to be $(+, 0, -, \ldots, -)$.

And this is precisely what we get from our computations. The essential results are shown in Fig. 11.6. The diagram shows the first three LCEs (the other seven are all strongly negative). One can notice that the convergence is quite strong and the exponent is distinctly positive (≈ 0.35).

The interpretation is that the motion of the system is strongly convergent towards the attractor from all directions but two. The zero Lyapunov exponent is associated with the direction of the motion along the flow. The presence of one positive exponent indicates that, on the attractor, there exists a direction along which nearby trajectories, on the average, diverge exponentially.

The fractal dimension has been computed (always for $r = 5$) by means of the Grassberger and Procaccia method. The results are shown in

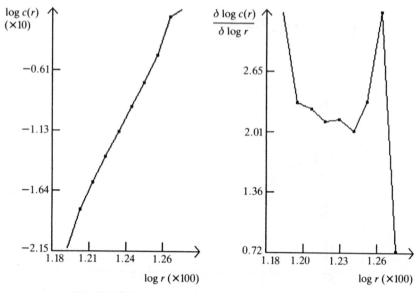

Fig. 11.7. Fractal dimension of the strange attractor.

Fig. 11.7.[13] The estimated dimension is ≈ 2.15 which is consistent with the geometrical aspect of the attractor. It is also close to the so-called Lyapunov dimension.[14] If we denote by χ_1, χ_2, χ_3 the first three LCEs, listed in descending order, we have in this case $\chi_1 \approx 0.35$, $\chi_2 \approx 0.00$, $\chi_3 \approx -4.2$. Consequently, we obtain:

$$D_L = 2 + \frac{\chi_1 + \chi_2}{|\chi_3|} \approx 2.08.$$

11.5 A RECONSTRUCTED 1D MAP

Recalling some earlier remarks concerning one-dimensional representations of large-dimensional systems (see Chapter 4), and considering that system (11.12) and (11.13) is characterized by strong dissipation,[15] one may wonder whether the essential features of its dynamics could be cap-

---

[13] Details of the construction of the two diagrams of Fig. 11.7 can be found in Chapter 7, section 7.2.

[14] See Chapter 7 for details.

[15] The rate of dissipation, or contraction, can be measured by the trace of the Jacobian matrix of the system, averaged on the attractor. In the present case, the trace is constant and equal to $-n^2$.

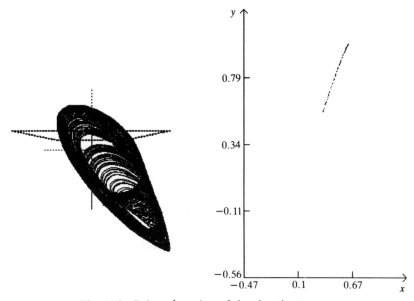

Fig. 11.8. Poincaré section of the chaotic attractor.

tured by means of a one-dimensional, discrete-time representation. After all, this was the starting point of our investigation.

In order to apply the '1D map approach' to our sysytem, we shall consider the chaotic case $r = 5$. Let us construct a (planar) Poincaré surface of section, transverse to a 3D projection of the attractor, and study the successive intersections of a trajectory with this surface. This is illustrated in Fig. 11.8. We observe that, if a suitable projection of the attractor is chosen and the surface is suitably positioned across the flow, the intersection looks like a collection of points lying on an open curve. This is an interesting result and, as we have already mentioned in Chapter 4, it provides in itself a symptom of chaotic behaviour.[16]

Next, let us measure the values of either of the co-ordinates (in the section plane) of each intersecting point, and let z_n denote the sequence (in time) of these values. Clearly, there are no *a priori* reasons for there to be any definite relation between such consecutive values, and, in general,

---

[16] Notice that, in principle, the choice of a particular cut transverse to the flow should not affect in any essential way the result we are aiming at. Indeed, taking a different cut corresponds to a co-ordinate transformation on the return map, and a theorem by Oseledec (1968) guarantees that the LCEs spectrum is invariant with respect to co-ordinate transformation. See on this point, Lichtenberg and Lieberman (1983, p. 412).

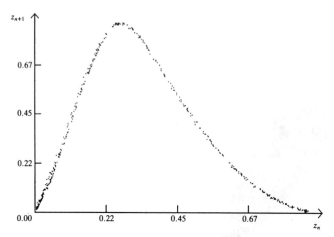

Fig. 11.9. The reconstructed one-dimensional map.

we would expect them to be scattered throughout the plane (within the limits of the intersecting curve). However, if we plot z_{n+1} as a function of its antecedent z_n, we obtain a well-defined curve, shown by Fig. 11.9, which has a familiar one-hump shape, with a quite steep slope.

To complete our experiment, we can try to estimate the (single) LCE associated with the reconstructed 1D map. As we know from the discussion of Chapter 6, the LCE in this case is defined by:

$$\chi = \lim_{n \to \infty} \frac{1}{n} \sum_{i=1}^{n} \ln|f'(z_i)|,$$

where $f'(z_i)$ is the derivative of the map at z_i. In practice, one takes n large enough to show a strong convergence of the LCE to a limit value. By fitting the data (i.e., the points of the first-return map) by cubic splines, it is possible to calculate the derivative at each point, and subsequently compute the LCE. The result we have obtained is $\chi \approx 0.68$, which indicates that the dynamics generated by the reconstructed 1D map are chaotic.

Notice an important point here. We have been able to show that an essential characteristic of the dynamics of our (continuous) multidimensional system can be captured by a one-dimensional discrete formalism. However, this does *not* mean that we can find a constant time delay such that a functional relationship can be defined between any two successive values of the representative variable. On the contrary, the time interval between each two successive intersections of the 'reconstructed' Poincaré map will in general be different. In this essential respect this map differs from (and is economically more satisfactory than) the fixed-delay map usually employed in the literature.

11.6 CONCLUSIONS

Altogether our findings seem to confirm the possibility of chaotic dynamics in continuous-time economic models, without the need to make recourse to the unduly strict and unrealistic assumptions concerning the lag structure.

Two considerations in particular seem to be prompted by our analysis. First of all, for the class of models under investigation, it is the combination of non-monotonic nonlinearities of the one-hump type *and* a certain structure of the lags that produces chaos. The extreme assumptions implicit in one-dimensional, fixed-delay models appear unnecessary. Multiple exponential lags possess a much greater flexibility, permitting one to model a large variety of economic situations, and their two controlling parameters, T and n, can be estimated econometrically.

On the other hand, the results of this chapter indicate that a sufficiently high order of the exponential lag, which corresponds to a sufficiently low variance in the distribution of individual reaction times, is a necessary (though not sufficient) condition for chaos to occur. We shall have more to say on the relationship between the pattern of the lags and the insurgence of chaos in Chapter 13 below.

12

Cycles and chaos in overlapping-generations models with production

12.1 THE BASIC MODEL

In the present chapter, we shall discuss a discrete-time dynamic model whose economic motivation can be found in some recent works on 'equilibrium cycles' in overlapping-generations models. In what follows we shall draw upon Reichlin's version (1986), which, in turn, is an extension of Grandmont's (1985).

Let us consider an economic system where events (production, consumption, etc.) take place at discrete points in time designated by $t \in \mathbb{N}^+$ and loosely referred to as 'periods',[1] and let us make the following assumptions:

(1) Population is stationary and consists of overlapping-generations of identical consumers living two periods, denoted respectively by the terms 'youth' and 'old age'.

(2) Consumers have perfect foresight about future prices and quantities.

(3) Consumption only occurs in old age and production (and saving) in youth.

(4) Consumers have a separable utility function whose arguments are consumption c and labour l.

(5) In each period t, a certain amount of the single homogeneous output x (say 'corn') is produced within a fixed coefficient technology, by means of inputs of current labour l, and capital ('seed corn') k saved and invested in the previous period.

Formally, we can write

$$x_t = \min(al_t, bk_{t-1}), \tag{12.1}$$

---

[1] We omit here all the problems arising from this description of economic activity. For example, it is not clear in such a setting when and how 'changes' in economic variables take place.

where a and b are the output/labour and the output/capital coefficients, respectively, and

$$k_t = x_t - c_t. \tag{12.2}$$

In order to ensure that the economy is 'viable' we shall assume that $b > 1$, i.e., the fixed requirement of seed corn per unit of harvest corn is less than one. Moreover, in order to economize in the use of parameters, we shall choose the unit of measure of labour so that $a = 1$.

If we denote the wage rate by w and the real interest rate by R, the zero-profit competitive hypothesis requires that we have

$$R_t = b[1 - w_t], \tag{12.3}$$

which is of course reminiscent of Sraffa's (1960) equation for the rate of profits.

Households are assumed to maximize utility at each point of time t by solving the following decision problem:[2]

$$\max_{l_t}\{u(c_{t+1}) - v(l_t)\}, \tag{12.4}$$

$$\text{s.t.} \quad k_t \leq w_t l_t$$
$$c_{t+1} \leq R_{t+1} k_t$$
$$c_{t+1}, \ l_t, \ k_t \ \geq \ 0,$$

where u and v are the two separable components of the utility function.

To tackle problem (12.4), we assume that the functions u, v are continuous on $[0, +\infty)$ and that for any $c, l > 0$, we have

$$u'(c) > 0, \qquad u''(c) < 0,$$
$$v'(l) > 0, \qquad v''(l) > 0.$$

Moreover,

$$\lim_{t \to \infty} v'(l) = \infty$$
$$\lim_{t \to 0} v'(l) = 0.$$

The necessary (and, under the postulated hypotheses, sufficient) condi-

---

[2] To make the problem manageable, an exception must be made for the first generation, which is assumed to be born at $t = 1$, and to be endowed with a given amount of corn entirely consumed in that period.

tions for a sequence $\{c, l\}_t$ to solve problem (12.4) are[3]

$$\mathscr{U}(c_{t+1}) = \mathscr{V}(l_t) \tag{12.5}$$

$$c_{t+1} = R_{t+1} l_t w_t \tag{12.6}$$

where $\mathscr{U}(c) = u'(c)c$ and $\mathscr{V}(l) = v'(l)l$. Whether conditions (12.5) and (12.6) can actually define forward discrete-time dynamics of the system or not, depends on the properties of the function $\mathscr{U}(c)$ and therefore on those of $u(c)$.

Two basic cases can be distinguished, corresponding to two types of instantaneous utility functions $u(c)$ frequently used in intertemporal optimizing models, namely:

(i)

$$u(c) = \frac{1}{\alpha} c^{\alpha}, \qquad 0 < \alpha < 1.$$

$u(c)$ is inelastic, i.e., it has a constant elasticity of substitution, equal to $1/(1 - \alpha)$. This case is also known as the 'constant relative risk aversion' (CRRA)[4] utility function, where the RRA coefficient is $-u''(c)c/u'(c) = (1 - \alpha)$.[5]

(ii)

$$u(c) = C - \frac{1}{\delta} r e^{-\delta c}, \qquad r > 0.$$

In this case, the elasticity of substitution is equal to $(\delta c)^{-1}$, i.e., it is decreasing in the level of consumption. As concerns the attitude to risk, this function is known as 'constant absolute risk aversion' (CARA), where the ARA coefficient is $-u''(c)/u'(c) = \delta$. In this case, saving may be a decreasing function of the rate of interest. Reichlin (1986) observes that, in overlapping-generations models without production, case (i) excludes the possibility of cycles or more complicated dynamic behaviours. He then proceeds to show that, when production is introduced into the picture, cyclical behaviour is possible even with a utility function of type

---

[3] The first order conditions can be written as

$$u'(c_{t+1})\frac{\partial c_{t+1}}{\partial l_t} - v'(l_t) = 0$$

$$c_{t+1} = w_t l_t R_{t+1}.$$

Substituting, we obtain $\mathscr{U}(c_{t+1}) - \mathscr{V}(l_t) = 0$.

[4] As has already been noticed (see Grandmont, 1985), to speak of 'risk aversion' in the context of a perfect foresight model is rather odd, but we shall retain this convenient characterization to detect different types of utility functions.

[5] It can also be shown that, in this case, there exists a positive functional relationship between saving and the rate of interest.

(i) . We shall provide an analytical-numerical analysis of both cases in turn, concentrating on the conditions for the occurrence of complex behaviour.

12.2 CONSTANT RELATIVE RISK AVERSION: FORWARD DYNAMICS

To analyse this case, let us postulate the following utility functions:

$$u(c) = \frac{1}{\alpha}c^{\alpha}, \qquad 0 < \alpha < 1,$$

and

$$v(l) = \frac{1}{\beta}l^{\beta} \qquad \beta > 1.$$

Consequently we have:

$$\mathcal{U}(c) = c^{\alpha}, \quad \mathcal{U}'(c) = \alpha c^{\alpha-1} > 0, \text{ for } c \in [0,\infty)$$

$$\mathcal{V}(l) = l^{\beta}, \quad \mathcal{V}'(l) = \beta l^{\beta-1} > 0, \text{ for } l \in [0,\infty).$$

In this case, clearly $\mathcal{U} : [0,\infty) \to [0,\infty)$, and $\mathcal{U}^{-1}$ is globally defined on $[0,\infty)$. By making use of equations (12.1) and (12.3), we can now write the following system of two first-order difference equations in c and l:

$$c_{t+1} = f(l_t), \tag{12.7}$$

$$l_{t+1} = b(l_t - c_t), \tag{12.8}$$

where $f(l) = \mathcal{U}^{-1}[\mathcal{V}(l)] = l^{\gamma}$, $\gamma = \beta/\alpha > 1$.

If we use the logarithmic scale, we could say that γ measures the ratio between the level of (expected) consumption 'tomorrow' required to induce consumers to perform a certain amount of labour 'today', and that amount of labour. For simplicity's sake, we shall henceforth refer to γ as the 'consumption/labour ratio'.

System (12.7)–(12.8) defines the future evolution of an economy starting at arbitrary initial conditions and satisfying the following requirements:

(i)   At each point in time consumers choose amounts of consumption and labour supply such that utility is maximized subject to the stated constraints.

(ii) Consumers perfectly foresee future levels of c and l, so that decisions taken 'today' in the expectation of a certain state of the economy 'tomorrow' are never proved wrong.

Let us proceed to investigate the dynamics of (12.7)–(12.8). Clearly, only two fixed points (equilibria) exist for this system, i.e.,

(1) $E_1 : \bar{c}_1 = \bar{l}_1 = 0;$

(2) E_2:

$$\bar{c}_2 = \left(1 - \frac{1}{b}\right)^{1/(\gamma-1)} > 0,$$

$$\bar{l}_2 = \left(1 - \frac{1}{b}\right)^{\gamma/(\gamma-1)} > 0,$$

for $b > 1$.

The Jacobian matrix of (12.7)–(12.8) calculated at the equilibrium point E_1 is equal to

$$J_1 = \begin{bmatrix} 0 & 0 \\ -b & b \end{bmatrix},$$

therefore tr $J_1 = b > 1$, det $J_1 = 0$.

At E_2 the Jacobian matrix is

$$J_2 = \begin{bmatrix} 0 & \gamma\left(1 - \frac{1}{b}\right) \\ -b & b \end{bmatrix},$$

and therefore tr $J_2 = b > 1$, det $J_2 = \gamma(b - 1) > 0$.

The necessary and sufficient conditions for local stability of system (12.6)–(12.7) can be written thus:[6]

(1) $1 - \text{tr } J + \det J > 0$;

(2) $\det J < 1$;   (12.9)

(3) $1 + \text{tr } J + \det J > 0$.

It can be seen immediately that, for equilibrium E_1, condition 1 is always violated for $b > 1$. Vice versa for equilibrium E_2, for $\gamma > 1$ and $b > 1$, conditions 1 and 3 always hold. In this case, therefore, local stability depends only on condition 2.

It follows that the only non-trivial equilibrium will be more likely to be unstable the greater the output/capital and the consumption/labour ratios. On the other hand, the matrix J_2 will have complex conjugate eigenvalues (which implies the presence of oscillations around the equilibrium point $(\bar{c}_2, \bar{l}_2)$), if and only if

$$(\text{tr } J_2)^2 < 4 \det J_2.$$ (12.10)

In order to illustrate how stability and oscillations depend on the system parameters γ and b, we shall define the following curves in the (γ, b) plane:

(1)   A 'divergence boundary' $\Delta_1(\gamma, b) = \det J_2 = \gamma(b - 1) = 1$, which

---

[6] Cf., for example, Thompson and Stewart (1986, p. 155).

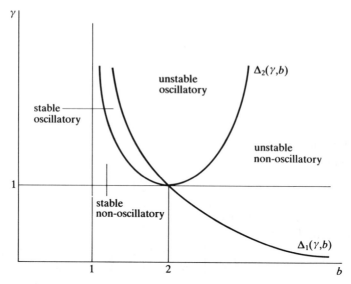

Fig. 12.1. The CRRA model: Hopf bifurcation boundary.

divides the relevant region of the plane into two subregions corresponding, respectively, to (locally) stable or unstable systems.

(2) A 'flutter boundary' $\Delta_2(\gamma, b) = (\mathrm{tr}\, J_2)^2 - 4\det J_2 = 0$, which divides the relevant region of the plane into two subregions corresponding, respectively, to (locally) oscillatory or non-oscillatory systems.

This is illustrated in Fig. 12.1.

From Fig. 12.1 it can be seen that, for values of b and γ sufficiently near 1, the equilibrium E_2 is stable without oscillations. Increasing γ or b (or both), the 'flutter boundary' will always be crossed before the 'divergence boundary' is reached. Then the equilibrium E_2 can only lose its stability through a Hopf (or Neimark) bifurcation. Indeed, for any value of $\gamma \in (1, +\infty)$, there always exists a value of $b \in (1, 2)$ producing such a bifurcation, and, vice versa, for any $b \in (1, 2)$ a Hopf bifurcation will occur for some $\gamma \in (1, +\infty)$.

As we have seen in Chapter 2, under certain generically verified conditions, a Hopf bifurcation for a system like (12.7)–(12.8) leads to the appearance of an invariant closed curve which may or may not be attractive, according to the sign of a certain coefficient which can be computed from the knowledge of the system. We have also seen that the dynamics of the system *on* the invariant curve may be periodic or quasiperiodic, according to whether a certain 'rotation number' (which can also be computed from the knowledge of the system) is rational or irrational. Finally, we have seen that when increasing (or decreasing) the control parameter beyond the bifurcation point, a transition to chaos may take place through one or the other of the routes discussed in Chapter 9, i.e., period-doubling, intermittency or blue sky catastrophe.

We shall not pursue the analysis of (12.7)–(12.8) any further here, but shall instead illustrate the situation by means of a few numerical simulations, performed fixing γ and varying the other parameter b. First of all, it can be shown that, for $\gamma = 6$, the equilibrium is stable for $b < 1.1\bar{6}$. Orbits will approach equilibrium without oscillations (for $\approx b \leq 1.15$) or with oscillations ($\approx 1.15 < b < 1.1\bar{6}$).

At $b = 1.1\bar{6}$ a Hopf–Neimark bifurcation occurs and an attracting in-variant circle appears. The motion on the circle is illustrated in Fig. 12.2a (quasiperiodic motion covering the entire circle).

Fig. 12.2b instead illustrates the phenomenon called 'mode-locking' (different frequencies get 'locked' together). In this case we have a period 5 cycle corresponding to five isolated points.[7]

12.3 CONSTANT ABSOLUTE RISK AVERSION: BACKWARD DYNAMICS

To analyse this case, we postulate the following utility function:[8]

$$u(c) = C - re^{-c}$$

where r is a parameter > 0[9] and C is a positive constant. Clearly, we have

$$u'(c) = re^{-c} > 0, \qquad u''(c) = -re^{-c} < 0,$$

and

$$\mathscr{U}(c) = u'(c)c = rce^{-c}, \qquad c \in [0, +\infty).$$

We shall, in contrast, retain the same function $v(l)$ employed in the analysis of case (i) above, namely:

$$v(l) = \frac{1}{\beta}l^\beta$$

---

[7] The simulation corresponding to Fig. 12.2b was performed for the following values of the parameters: $\gamma = 36; b = 1.051$.

[8] For simplicity's sake, we have used the function of case (ii) above (section 12.1) with $\delta = 1$.

[9] Notice that r can be interpreted as the value of marginal utility of consumption when $c = 0$. Clearly, for this consumption function there exists a maximum value of utility (a 'bliss') equal to $u(\infty) = C$.

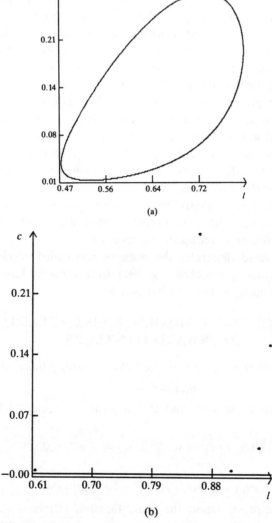

Fig. 12.2. The CRRA model: (a) the invariant circle; (b) mode-locking.

The reader will readily verify that now the function $\mathscr{U}(c)$ has a 'one-hump' form, and does not admit a global inverse $\mathscr{U}^{-1}(c)$. Therefore, in this case we cannot define forward dynamics as we did in the previous case ((12.7)–(12.8)). Since $\mathscr{V}(l)$ is invertible, however, we could write the system of equations:

$$l_t = g(c_{t+1}) \qquad (12.11)$$

$$c_t = g(c_{t+1}) - \frac{1}{b} l_{t+1}, \qquad (12.12)$$

where $g(c) = \mathscr{V}^{-1}[\mathscr{U}(c)] = [rce^{-c}]^{\frac{1}{b}}$.

Given the initial conditions for l and c, system (12.11)–(12.12) determines their dynamics 'in the past'.

Adopting (and keeping in mind) the convention that successive (integer) values of the independent variable t, $1, 2, \ldots$ are now taken to correspond to points on the negative half of the real line, we can rewrite (12.11)–(12.12) thus:

$$l_{t+1} = g(c_t), \qquad (12.13)$$

$$c_{t+1} = g(c_t) - \frac{1}{b} l_t, \qquad (12.14)$$

which is perfectly well defined for $c \in [0, +\infty)$ and $l \in [0, +\infty)$, $t \in \mathbb{N}^+$.

This is very much the same procedure followed by Grandmont (1985). Naturally, one may wonder what economic meaning can be attributed to a solution of system (12.13)–(12.14). A 'minimal' interpretation is as follows. A sequence of values of l and c generated by (12.13)–(12.14) over an interval, say, $t \in (1, T)$, corresponds to values of labour and consumption from 'now' to a certain date in the past. When read 'forward', i.e., from T to 1, the same sequence can be interpreted as a succession of values, and therefore, in a sense, dynamics of those variables, which, over the interval considered, satisfy the criteria both of optimality at each instant of time and of perfect foresight.

Stronger, more interesting results might be obtained by using the backward dynamics described by (12.13)–(12.14) in order to extract information about 'true' forward dynamics appropriately defined. In his article, Grandmont (1985) establishes certain sufficient conditions ensuring that (local) stability of periodic orbits of the backward dynamics implies that of periodic orbits of 'true dynamics'.[10]

The latter is defined in terms of a certain expectations function which determines the expected values at time t as a function of past values over a given constant interval T. Under certain assumptions, the expectations are consistent with periodicity, i.e., agents make correct forecasts along a periodic orbit and small errors are followed by stabilizing adjustments which re-establish (periodic) perfect foresight dynamics.

This is an interesting result, but a number of questions may be raised. First of all, the assumptions under which the result holds should be given

---

[10] But not vice versa.

a stronger economic justification. Secondly, and more importantly, the uniqueness of the stable periodic orbit for unimodal maps with a negative Schwarzian derivative does not extend to higher order maps. Even two-dimensional discrete-time models, like the one studied in this chapter, may have an indefinitely large number of stable periodic orbits, each of which may consequently have a negligibly small basin of attraction. In this case, the significance of local stability may be called into question. The situation is even less clear in the case of aperiodic or chaotic dynamics, for which the definition of appropriate expectation functions has yet to be discovered. While awaiting further progress in this direction, the study of the backward dynamics described by (12.13)–(12.14) seemed to be a useful preliminary step, not least because it provides us with the opportunity to apply the techniques presented in the previous chapters to a relatively lesser known case.

From (12.13)–(12.14) we deduce that the system in question has two fixed (equilibrium) points, namely:

(i) E_1: $(\bar{c}_1, \bar{l}_1) = (0, 0)$;

(ii) E_2: $(\bar{c}_2, \bar{l}_2) = \{c, l : f_1(c) = f_2(c) \text{ and } l = [b/(b-1)]c\}$,
      where

$$f_1(c) = \left[\frac{b}{(b-1)r^{(1/\beta)}} \right] c^{1-1/\beta},$$

$$f_2(c) = e^{-c/\beta}.$$

It can easily be verified that, under the stated assumptions on the parameters b, β and r, $\bar{c}_2$ and $\bar{l}_2$ are both positive (Fig. 12.3a). The equilibrium E_1 is always unstable. Local stability and bifurcation of the unique non-trivial equilibrium, E_2, can be evaluated even without knowing the exact values of $(\bar{c}_2, \bar{l}_2)$. For this purpose, we shall consider $(1 < b < \infty)$ and $(1 < \beta < \infty)$ as constants and shall take $r \in (0, \infty)$ as a bifurcating parameter.

First of all, consider that, from the condition $f_1(c) = f_2(c)$, we can derive the function $r = r(\bar{c}_2)$, namely

$$r = e^{\bar{c}_2} \bar{c}_2^{\beta-1} \left(\frac{b}{b-1} \right)^{\beta},$$

whose general shape is illustrated by Fig. 12.3b.[11]

---

[11] Notice that, for all values of $b, \beta > 1$, $\lim_{\bar{c}_2 \to 0} r(\bar{c}_2) = 0$, $\lim_{\bar{c}_2 \to \infty} r(\bar{c}_2) = \infty$, $r'(\bar{c}_2) > 0$ for $\bar{c}_2 \in (0, +\infty)$, $\lim_{\bar{c}_2 \to \infty} r'(\bar{c}_2) = \infty$. The curve of Fig. 12.3b has been drawn with $\beta = 2$.

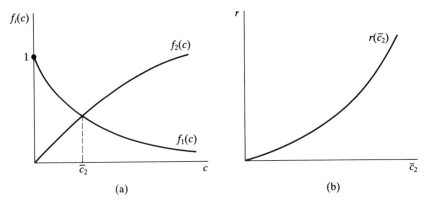

Fig. 12.3. Equilibrium conditions for the CARA model.

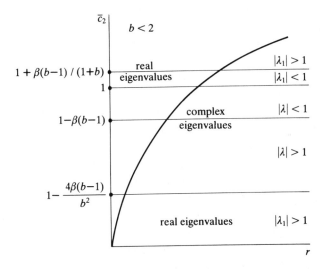

Fig. 12.4. Bifurcation boundaries for the 'low productivity' (LP) case.

Secondly, consider that the Jacobian matrix J_2 evaluated at the equilibrium $(\bar{c}_2, \bar{l}_2)$, is equal to:

$$\begin{bmatrix} 0 & \frac{b}{\beta(b-1)}(1 - \bar{c}_2(r)) \\ -\frac{1}{b} & \frac{b}{\beta(b-1)}(1 - \bar{c}_2(r)) \end{bmatrix},$$

whence

$$\text{tr } J_2 = \frac{b}{\beta(b-1)}(1 - \bar{c}_2(r)),$$

$$\det J_2 = \frac{1}{\beta(b-1)}(1 - \bar{c}_2(r)).$$

Recalling now the three (local) stability conditions, as well as the 'flutter' condition mentioned above, we can write:

(i) $1 - \text{tr } J_2 + \det J_2 > 0$ is always verified for $\beta > 1$ and $\bar{c}_2 > 0$.

(ii) $1 + \text{tr } J_2 + \det J_2 \gtreqless 0$ if $\bar{c}_2 \gtreqless 1 + \beta(b-1)/(1+b)$.

(iii) $\det J_2 \gtreqless 1$ if $\bar{c}_2 \gtreqless 1 - \beta(b-1)$.

(iv) $(\text{tr } J_2)^2 - 4\det J_2 \gtreqless 0$ if $\bar{c}_2 \gtreqless 1 - 4\beta(b-1)/b^2$.

Two basic situations may arise according to whether $1 < b < 2$ or $b > 2$. For convenience's sake, we shall label the former case 'low productivity', and the latter 'high productivity'. The former case is summarized by Fig. 12.4 (λ_1 indicates the largest eigenvalue in modulus).[12]

Here the equilibrium is stable for intermediate values of r and unstable for larger or smaller values. Starting from $(r = r(\bar{c}_2); \bar{c}_2 = 1)$ and *decreasing* it, we have a loss of stability through a Hopf (or Neimark) bifurcation at $(r = r(\bar{c}_2); \bar{c}_2 = 1 - \beta(b-1))$ (assuming that the latter expression is positive). As we know (see Chapter 9 above), under certain conditions, this leads to the appearance of an invariant circle on which the motion of the system can be periodic or quasiperiodic. The invariant circle is shown by Fig. 12.5.[13]

The practical interest of this occurrence is limited, however, by the fact that the circle is extremely small, it is not robust with regards to very small changes of r and, moreover, it has a very small basin of attraction.

A more interesting situation arises when r is increased from $(r(\bar{c}_2), \bar{c}_2 = 1)$ to $(r(\bar{c}_2), \bar{c}_2 = 1 + \beta(b-1)/(1+b))$. Here we have a loss of stability through a flip bifurcation. Numerical simulations, performed for values of $\beta = 2$ and $b = 1.55$, indicate that if we increase r beyond the first flip (at $r \approx 47.55$), the system undergoes a period-doubling sequence of bifurcations, eventually ending in a chaotic zone interspersed with periodic windows that is very similar to the one generated by the familiar one-dimensional unimodal maps (cf. Chapter 9). Fig. 12.6 illustrates the

---

[12] Notice that the expressions $1 - \beta(b-1)$ and $[1 - 4\beta(b-1)/b^2]$ may be negative and, if so, the corresponding boundaries become irrelevant.

[13] This simulation has used the following values of the parameters: $\beta = 2$; $b = 1.25$; $r \approx 20.60$.

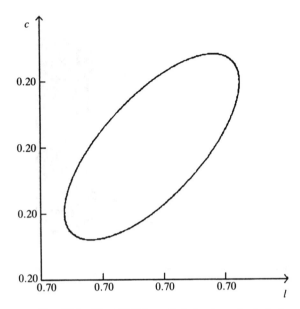

Fig. 12.5. The CARA model: the invariant circle.

situation by means of a bifurcation diagram. (Notice that, beyond a certain value of r, the attracting set ceases to be attracting and the system solutions become unbounded.)

To complete the description of this 'route to chaos', we have investigated the evolution of the power spectrum as r is increased. The results are shown in Fig. 12.7.

The progressive evolution from periodic behaviour, characterized by sharp distinct peaks at the leading frequencies, to chaotic behaviour (broad band power spectrum) is evident.

Finally we have picked up a value of r (i.e., $r = 80$) to which there seemed to correspond chaotic behaviour and we have tried to provide some qualitative and quantitative characterizations of the asymptotic dynamics of the system. The attractor shown in Fig. 12.8 exhibits the typical features attributed to strange attractors. In particular, the fractal structure of the attractor is quite impressively shown in the enlargements, which are reminiscent of those of the well-known Hénon attractor.

The estimates of the Grassberger and Procaccia dimension (≈ 1.44, see Fig. 12.9), although the convergence is not particularly good, seem to confirm that the attractor has indeed a non-integer dimension.

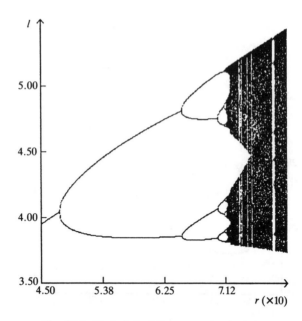

Fig. 12.6. Period-doubling route in the LP case.

Finally, the estimates of the (two) Lyapunov characteristic exponents (shown in Fig. 12.10) indicate that one of them is positive, and therefore the system possesses SDIC.

For the 'high productivity' case ($b > 2$) the bifurcation scenario is summarized by Fig. 12.11. In this case, for $\bar{c}_2 > 0$, $\det J_2 < 1$ always, and the 'flutter boundary' may or may not correspond to positive values of $\bar{c}_2$. At any rate, no Hopf bifurcation may occur. The loss of stability of the equilibrium takes place through a flip bifurcation ($\lambda_1 = -1$). This leads to a stable period 2 cycle.

Numerical investigation (for $b = 5$, $\beta = 2$) of this situation has given very interesting results. Increasing r beyond the first flip bifurcation (at $r \approx 37.57$), we find again a period-doubling sequence leading to chaotic behaviour. Numerical evidence is qualitatively similar to that obtained in the 'low productivity' case and the graphical results shown in Fig. 12.12 (bifurcation diagram) and Fig. 12.13 (power spectrum) do not need any detailed comments.

Fixing r within the chaotic zone (at $r = 105$), we have reproduced some enlargements of the attractor (see Fig. 12.14). Once again its fractal, Hénon-type features are very evident. The estimate of the correlation

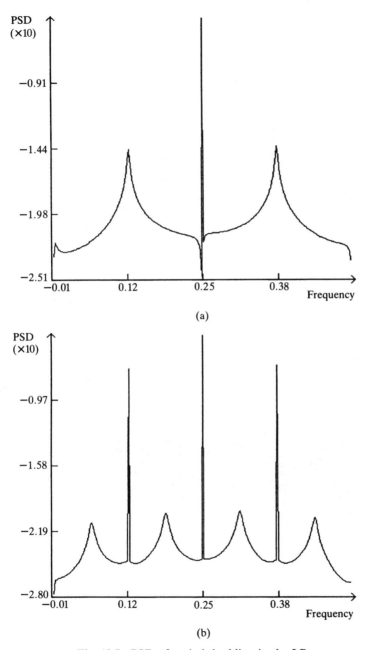

Fig. 12.7. PSD of period-doubling in the LP case.

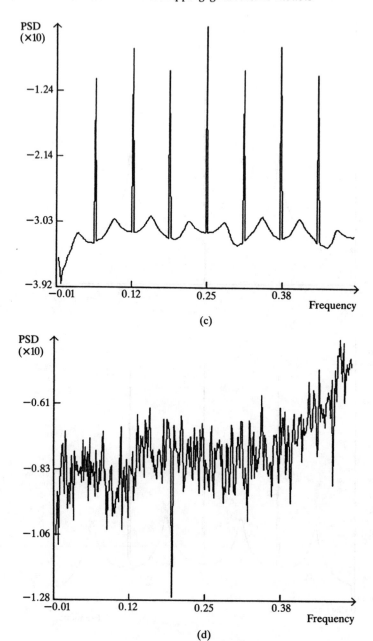

(c)

(d)

Fig. 12.7. (*cont.*)

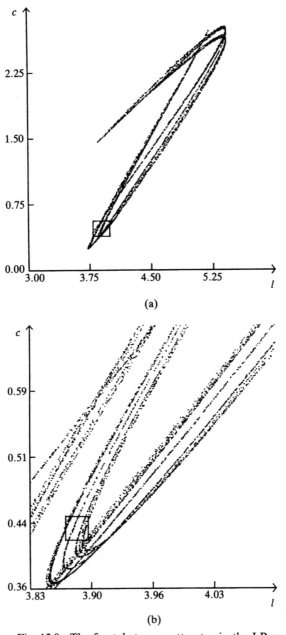

Fig. 12.8. The fractal strange attractor in the LP case.

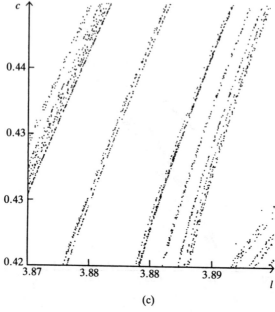

(c)

Fig. 12.8. (*cont.*)

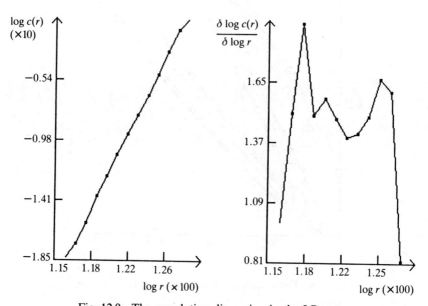

Fig. 12.9. The correlation dimension in the LP case.

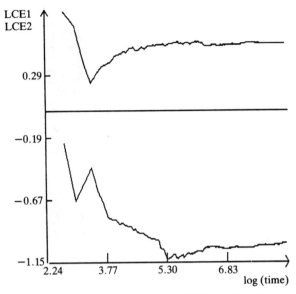

Fig. 12.10. The dominant LCEs in the LP case.

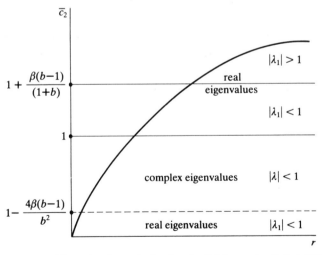

Fig. 12.11. Bifurcation boundaries for the 'high productivity' (HP) case.

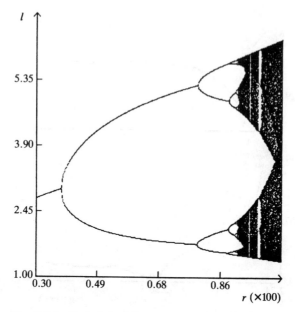

Fig. 12.12. Period-doubling route to chaos in the HP case.

dimension is also non-integer (≈ 1.12) and it is graphically illustrated in Fig. 12.15.[14] Finally, Fig. 12.16 shows the LCE spectrum indicating that the dominant exponent is converging to a positive value.

In conclusion, the exercises performed in this chapter have shown that the solutions of an overlapping-generations model with production may be more complex than simple periodic, or quasiperiodic orbits. Insofar as our numerical simulations can be relied upon, they strongly suggest that those solutions may indeed be chaotic.

---

[14] Unfortunately, the convergence is again less than satisfactory.

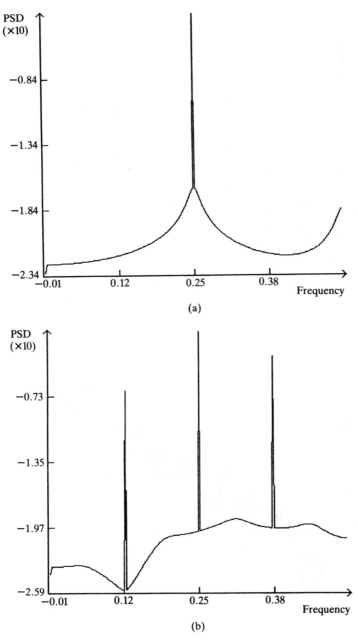

Fig. 12.13. PSD for the HP case.

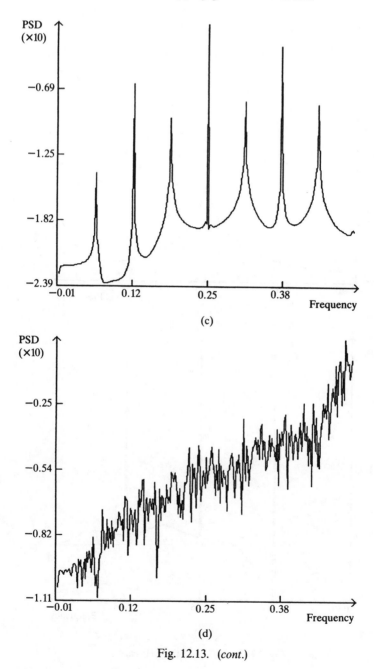

Fig. 12.13. (*cont.*)

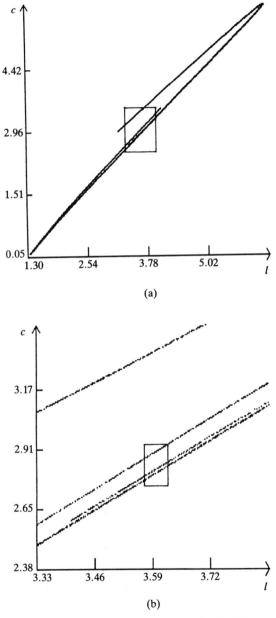

(a)

(b)

Fig. 12.14. The strange attractor in the HP case.

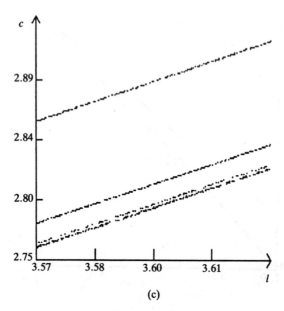

(c)

Fig. 12.14. (cont.)

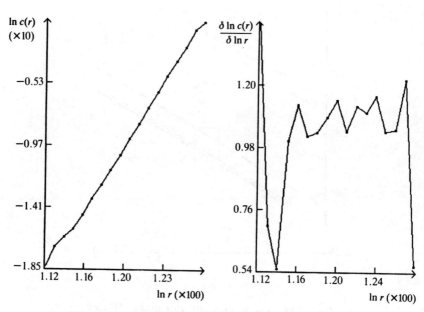

Fig. 12.15. The correlation dimension in the HP case.

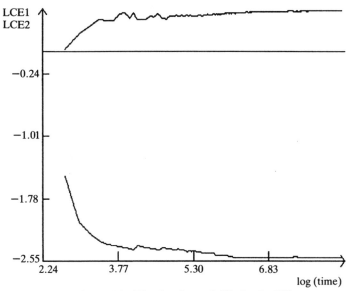

Fig. 12.16. The dominant LCEs in the HP case.

13

Chaos in a continuous-time model of inventory business cycles

13.1 THE MODEL

The mechanisms which produce deterministic periodic or chaotic dynamics can often be reduced to combinations of lags and nonlinearities. In Chapter 11 we investigated a particularly important class of those combinations involving a continuous-time multiple exponential lag (MEL) and a single nonlinearity of a 'one-hump' type. We showed that chaotic dynamics may only occur in this case if the order of the lag n is rather large.[1] When the lag structure and/or the (nonlinear) input to it are of a different kind, however, complicated dynamics may occur at a much lower order of the lag.[2]

To see this we shall investigate a continuous-time model whose economic motivation can be found in the early analysis of inventory business cycles (Metzler, 1941), and, more recently, in the works of Gandolfo (1983) and Lorenz (1989b). The latter author was the first to point out the relation between the economic model in question and certain mathematical results of Arneodo et al. (see, for example, 1981, 1982), concerning the so-called 'spiral chaos'.

Our own model closely follows Lorenz's, but we shall manipulate it somewhat with a view to establishing a relation with the previous discussion of lags. Let us define the following notation:

Y = actual net national income (output)
B = inventory stock
S = global (net) saving

---

[1] We have seen that, with a nonlinearity of the logistic type, chaos may occur when $n \approx 10$. If the nonlinearity is of an exponential type (i.e., $f(x) = rxe^{-x}$), a much greater order ($n \approx 50$) is required (cf. Sparrow, 1980; Invernizzi and Medio, 1991).

[2] Of course, three is the minimum order of a system of o.d.e. for which complex behaviour may occur anyway.

I = global (net) investment

b = desired stock/output ratio, constant.

Let the index e denote the expected value of the relevant variable. Suppose now that output adjusts in response to discrepancies between desired and actual inventory stock through a first-order exponential lag, thus:

$$T\dot{Y} = (bY^e - B), \qquad T > 0 \tag{13.1}$$

T being the inverse of the speed of adjustment or, in other words, the 'length' of the lag.

Suppose further that expected output is a linear function of the level, velocity and acceleration of actual output,[3] thus:

$$Y^e = G_1(D)Y \tag{13.2}$$

where $G_1(D) = a_2D^2 + a_1D + 1$, $D = d/dt$, and a_1 and a_2 are positive constants.

The increase in inventory is equal to the excess of (net) saving over (net) investment, i.e.,

$$\dot{B} = S - I. \tag{13.3}$$

Finally, we shall adopt Kaldorian saving and investment functions (cf. Kaldor, 1940, pp. 78–92) as depicted in Fig. 13.1.

The original (verbal) formulation of Kaldor seems to assume a simple adaptive mechanism such as:

$$\theta\dot{Y} = [I(Y) - S(Y)] \quad \theta > 0,$$

which, in turn, implies that $\bar{Y}_1$ and $\bar{Y}_3$ are stable equilibria and $\bar{Y}_2$ is an unstable one. In Lorenz's (and our own) model, owing to the different specification of the adjustment mechanism, both the equilibria, $\bar{Y}_1$ and $\bar{Y}_2$, may be simultaneously unstable. This model may in fact be interpreted as an investigation of the dynamics of the economy when output is 'trapped' at, or in the vicinity of, the low-level interval $(\bar{Y}_1, \bar{Y}_2)$. It can easily be seen that the function:

$$F(Y) \equiv S(Y) - I(Y),$$

belongs to the unimodal class and can be formulated, for example, as:

$$F(Y) = mY(1 - Y). \tag{13.4}$$

---

[3] Notice that (13.2) may be interpreted as a simplified version of the hypothesis that agents' expectations on income are positively affected by the level, the rate of growth and the changes in the rate of growth of income. If the rate of growth of income (and its rate of change) are small, the two hypotheses are roughly equivalent.

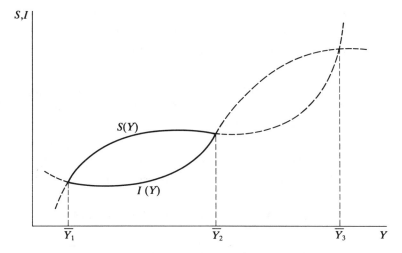

Fig. 13.1. Kaldorian saving and investment functions.

Combining equations (13.1)–(13.4) and putting $\bar{Y}_1 = 0$ and $\bar{Y}_2 = 1$[4] we can write:

$$G_2(D)Y = \hat{F}(Y) \tag{13.5}$$

where

$$G_2(D) = D^3 + c_2 D^2 + c_1 D,$$

$$c_1 = \frac{1}{a_2} > 0$$

$$c_2 = \frac{ba_1 - T}{ba_2} \gtreqless 0$$

$$\hat{F}(Y) = rY(1 - Y),$$

$$r = \frac{m}{ba_2} > 0.$$

Under the postulated assumptions, the Lie derivative (the divergence) of (13.5) is constant and equal to $-c_2$. Since we are interested here in the study of dissipative systems we shall henceforth only consider the case $c_2 > 0$.[5] System (13.5) has two equilibrium points: E_1, located at $Y = 0$, and E_2, located at $Y = 1$. It can be easily seen that E_1 is always

---

[4] Notice that in this case $Y < 0$ does not necessarily mean that output is negative.

[5] On conservative and dissipative systems, see our comments in Chapter 2 above. The reader can verify that, by putting $c_2 = 0$, one can obtain a conservative system with three Lyapunov characteristic exponents equal to zero.

unstable[6], whereas E_2 is stable if $r < c_1c_2$. We can readily establish that, if $c_2 > 0$, the Jacobian matrix calculated at E_2 cannot have any positive real root and one of the following situations occurs:

(i)　there are three negative real roots;

(ii)　there is one negative real root and a pair of complex conjugate roots[7] whose real part is greater or smaller than zero according to whether c_1c_2 is smaller or greater than r.

It follows that, when c_1c_2 becomes smaller than r, the system loses its stability through a Hopf bifurcation. Consequently, in this case, the loss of stability leads to the appearance of periodic solutions.

13.2　LAG STRUCTURE AND COMPLEX BEHAVIOUR

Arneodo *et al.* (1981, 1982), have proved that, for certain values of the parameters, system (13.5) satisfies the conditions of the Šilnikov theorem (see Chapter 9 above) and therefore possesses a horseshoe-like (chaotic) invariant set.[8] These authors' simulations also indicate that the chaotic set is attracting.

These results suggest the following interesting observation:

There exist low-order lag structures (e.g., the polynomial $G_2(D)$), different from an MEL, such that, when coupled with a nonlinearity of a 'one-hump' type, their interaction results in chaotic dynamics. In this case, chaos may occur at values of the order of the lag and/or of the steepness of the 'hump' much lower than would be necessary for an MEL.

It would be a promising research project to classify different classes of lags which, in combination with given classes of nonlinearities, may

---

[6] The reader can verify that, at E_1, the third Hurwitzian determinant Δ_3 is equal to $-\Delta_2 r$, Δ_2 being the second Hurwitzian determinant. Consequently, since $r > 0$, either Δ_2 or Δ_3 must be negative, which implies instability.

[7] Notice that complex roots will occur whenever c_2 is sufficiently small relative to c_1, independently of r. To see this, consider the auxiliary equation of (13.5) at E_2, i.e.,

$$\lambda^2 + c_2\lambda_2 + c_1\lambda + r = 0. \qquad (*)$$

In any algebra textbook dealing with third degree equations, one can verify that a sufficient condition for (*) to have two complex roots is that

$$q = \frac{1}{3}c_1 - \frac{1}{9}c_2^2 > 0,$$

which clearly obtains if c_2 is small relative to c_1.

[8] In fact, Arneodo *et al.* (1982), adopting a semilinear approximation of $F(Y)$, proved *analytically* the presence of a homoclinic orbit and of the other conditions of the Šilnikov theorem.

produce complex behaviour. Such an ambitious goal could not be pursued here, so we shall limit ourselves to addressing some specific questions concerning the present model.

First of all, let us pose the following question: what makes the lag mechanism $G_2(D)$ capable of producing a chaotic output, when it is coupled to a function like (13.4)? We do not have a fully developed mathematical answer, but we shall make a conjecture. The model under discussion can be described in terms of a closed, single-loop feedback system, which consists of a lag mechanism (the polynomial $G_2(D)$), and a nonlinear function $\hat{F}(Y)$, and is similar to that illustrated in Fig. 11.1.

In Chapter 11, we have seen that, under certain conditions, a correspondence can be established between a lag operator of a dynamical system and a 'weighting function' representing the impact on current output of inputs delivered at different times in the past. We have also seen that, when the operator takes the form of an MEL, this weighting function can be interpreted as a probabilistic version of a one-dimensional difference equation.

Consider now the lag operator $G_2(D)$. As we know from Chapter 11, the corresponding weighting function $w(t)$ can be found by means of the Laplace transform (LT) mechanism, as follows:

$$w(t) = \mathcal{L}^{-1}[1/G_2(p)],$$

where p as usual indicates the complex variable of the transform and $\mathcal{L}^{-1}$ is the inverse of the LT operator. Let λ now be a root of the polynomial $G_2(D)$, and let us consider that Arneodo *et al.* found chaotic behaviour of system (13.5) for values of the parameters $c_1 = 1$, $0.4 < r < 1$ and $c_2 \approx 0.4$. To those values of the parameters c_1 and c_2, there correspond one zero and two complex conjugate roots with negative real parts.[9]

The weighting function $w(t)$ of the lag operator $G_2(D)$ in this case will have the general form:

$$w(t) = \frac{1}{c_1} + Ae^{-\sigma t}\cos(\omega t + \theta), \tag{13.6}$$

where we have put $\lambda = -\sigma \pm i\omega$, $\sigma > 0$, and A and θ are real numbers which depend only on the parameters. The weighting function corresponding to the lag $G_2(D)$, therefore, is a damped sinusoid with frequency ω.[10]

If we compare (13.6) with the weighting function associated with an

---

[9] In the following discussion, the reader should keep in mind the distinction between the roots of the lag polynomial and the roots of the auxiliary equation of the entire system.

[10] Cf. Doetsch (1971, ch. 3).

MEL, i.e.,

$$w(t) = \left(\frac{n}{\tau}\right)^n \frac{t^{n-1}}{(n-1)!} e^{-nt/\tau},$$

(cf. Chapter 11, equation (11.8)), we observe that, whereas the latter is *unimodal* for any $n > 2$, the weighting function of $G_2(D)$ is multimodal whenever its roots are complex. It is this structure of the time profile of the lag, we believe, that, when coupled to a nonlinear input function of a 'one-hump' kind like $\hat{F}(Y)$, can produce a chaotic output. A hint for understanding this fact (although by no means a rigorous explanation) may be found in the following considerations.

Taking the same approach as discussed in Chapter 11, equation (13.6) may be viewed as a probabilistic representation of an infinite dimensional, discrete-time lag. However, if the damping factor σ is sufficiently large, all but the first few dominant lags (corresponding to the dominant modes) can be neglected. The following simple example involving only two lags will help clarify the issue.

Suppose we have a discrete-time dynamical system such that a linear relation exists between the value of a variable Z at time t and the values of another variable Y at times $t + \tau_1, t + \tau_2$ $(\tau_1, \tau_2 > 0)$. In general, Z may be a (linear or nonlinear) function of Y. We can then write:

$$b_2 Y_{t+\tau_2} + b_1 Y_{t+\tau_1} = Z_t, \qquad (13.7)$$

where $b_i > 0$ and constant for $i = 1, 2$. Recalling our discussion of Chapter 11, a continuous-time generalization of (13.7) can be written as

$$G_*(D) Y = Z \qquad (13.8)$$

where

$$G_*(D) = \left\{ b_2 \left(\frac{\tau_2 D}{n} + 1\right)^n + b_1 \left(\frac{\tau_1 D}{n} + 1\right)^n \right\}$$

A possible (but by no means the only) interpretation of (13.8) is that it refers to an economy characterized by two classes of agents, each comprising an indefinitely large number of members, and differing from one another in their speeds of reaction to economic stimuli. Each class is characterized by a different expected value τ_i $(i = 1, 2)$ of the overall reaction time and by a variance which represents the 'spread' of individual reaction times of the members of each class around the mean. (For simplicity's sake, we have assumed here that the two classes have the same variance.) We know that, as $n \to \infty$, (13.8) $\to$ (13.7) and we are back to the case of two fixed delays. This corresponds to the situation in which the agents belonging to each class are perfectly homogeneous, whereas the two classes are different from one another. By putting $G_*(D) = 0$

and solving for D, it is easy to see that, the polynomial $G_*(D)$ has at most one (negative) real root, all the others being complex conjugate pairs. Therefore, for $n \geq 2$, the corresponding 'weighting function' is multimodal and it tends to *two* Dirac delta functions situated at τ_1 and τ_2, as $n \rightarrow +\infty$.

Therefore, whereas an MEL can be seen as a continuous-time generalization of a discrete-time lag of order one, the lag operator $G_2(D)$ can be viewed as a similar generalization of discrete-time lags of order two or higher. Now, as we have mentioned in Chapter 1 of this book, the potential complexity of nonlinear maps is known to undergo a 'leap' whenever their dimension is increased by one unit. Then, if our reasoning so far is correct, it is little wonder that the lag operator $G_2(D)$ may generate more complex dynamic behaviour than does an MEL of the same order, coupled to the same nonlinearity.

13.3 PARAMETER ANALYSIS AND NUMERICAL SIMULATIONS

In order to explore this point and seek confirmation of our hypothesis, let us go back to system (13.5) and perform some parametric analysis. There are three parameters in the system: c_1, c_2 and r. Recalling that the condition for instability of (13.5) is $c_1 c_2 < r$ and the condition for complex roots of the lag polynomial is $c_2^2 < 4c_1$, we can combine them as indicated in the diagrams of Fig. 13.2.

These diagrams can be used as a practical tool to locate zones of complex behaviour in the three-dimensional parameter space. Consider, for example, Fig. 13.2a. If the values of the parameters are so chosen that the system is positioned in the lower part of the stable zone, and, keeping c_2 and r fixed, c_1 is progressively decreased, we expect the system to go through a Hopf bifurcation and, possibly, a period-doubling transition to chaos. This conjecture is supported by the bifurcation diagram shown in Fig. 13.3.

The period-doubling sequence is quite evident. There is no sign of periodic windows, however, but this could be the consequence of the coarseness of the numerical simulation and its graphical representation. For low values of c_1 (≈ 1), there seems to be a sudden increase in the size of the chaotic attractor. This suggests the presence of the so-called 'interior crisis', or 'interior catastrophe', a phenomenon whose occurrence has also been detected for the logistic map. For even lower values of c_1, the system 'explodes' and its solutions become unbounded.

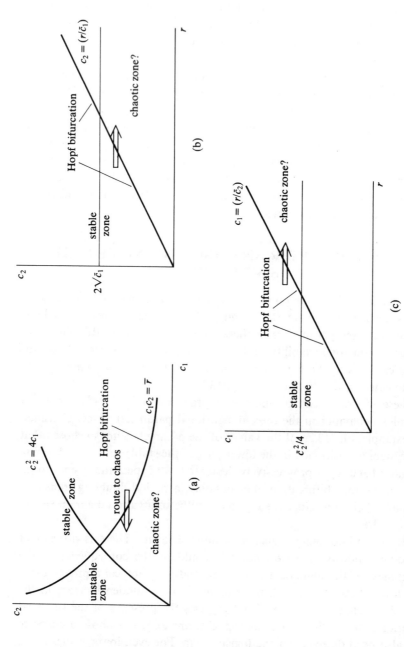

Fig. 13.2. Zones of different long-term behaviour of system (11.5): (a) in the (c_2, c_1) plane (with fixed $r = \bar{r}$); (b) in the (c_2, r) plane (with fixed $c_1 = \bar{c}_1$); (c) in the (c_1, r) plane (with fixed $c_2 = \bar{c}_2$).

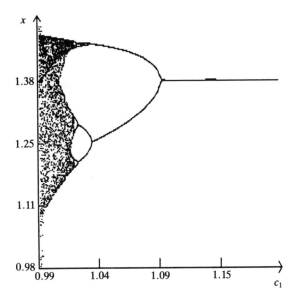

Fig. 13.3. Period-doubling route to chaos 1.

A similar procedure can be followed with regard to the diagram of Fig. 13.2b, appropriately fixing c_1 and c_2 and increasing r. The resulting bifurcation diagram is shown in Fig. 13.4.[11]

The results are qualitatively the same as in the previous experiment: a period-doubling route to chaos terminating in an 'explosion' of the system; no clear sign of periodic windows in the chaotic zone; some indication of an 'interior crisis' immediately before the 'explosion'.

The period-doubling route to chaos can be further appreciated by considering the diagrams of Fig. 13.5 (using the same values of c_1 and c_2), which show the final trajectories of equations (13.5) for different values of the controlling parameters. The changes in the behaviour of the system can also be detected by power spectrum analysis, as illustrated in Fig. 13.6, where the evidence of an evolution from periodic to aperiodic chaotic behaviour is quite strong. As we increase the control parameter r, we see peaks appear in the power spectrum corresponding to sub-multiples of the fundamental frequency. For even higher values of r, a 'noise floor' associated with chaotic behaviour appears and successively

---

[11] In order not to overburden the presentation, we shall omit the discussion of the diagram of Fig. 13.2c, which gives entirely similar results.

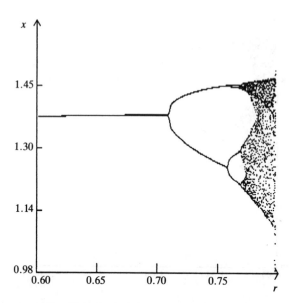

Fig. 13.4. Period-doubling route to chaos 2.

destroys the subharmonics in the reverse order of their appearance. Notice, however, that, insofar as our numerical analysis is correct, the 'noise floor' does not seem to destroy the peaks corresponding to the fundamental frequency (and some of its harmonics). This type of chaos, characterized by the co-existence in the power spectrum of broad band components and sharp peaks is sometimes called 'non-mixing chaos'. An example of non-mixing chaos is given by the Rössler attractor.[12]

Finally, we have estimated the LCEs and the (Grassberger and Procaccia) fractal dimension for the chaotic attractor corresponding to $c_1 = 0.99$, $c_2 = 0.4$, $r = 0.8$. The results are shown in Figs. 13.7 and 13.8. The dominant LCE is small but positive (≈ 0.08) and the convergence is good. The estimated fractal dimension is equal to ≈ 2.1, which is in harmony with the other qualitative information we have on the attractor. Altogether, our exercises strongly suggest that, for wide zones of the parameter space a 'multimodal' structure of the lag operator $G_2(D)$, associated with a 'one-hump' nonlinearity in the excess saving function, may generate cycles and chaos in the dynamics of the system.

---

[12] See Rössler (1976) and Oono and Osikawa (1980).

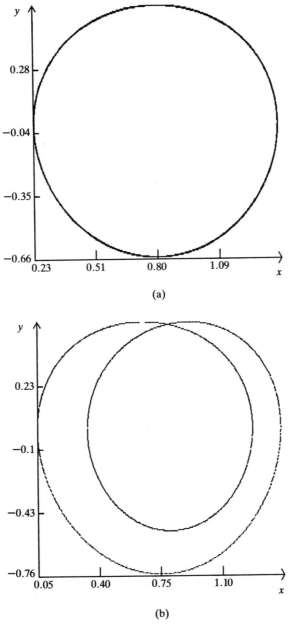

(a)

(b)

Fig. 13.5. Final trajectories of system (13.5) for different values of *r*.

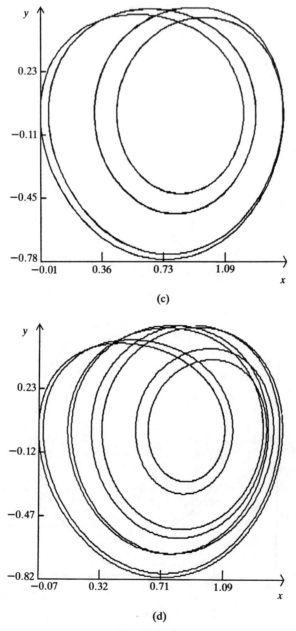

(c)

(d)

Fig. 13.5. (*cont.*)

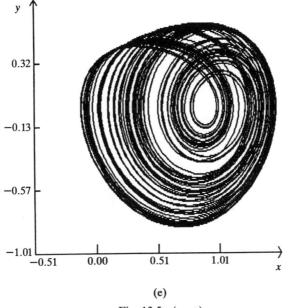

(e)

Fig. 13.5. (*cont.*)

13.4 A FEW COMPLICATIONS AND CONCLUSIONS

The constraints put on the lag structure (the polynomial $G_2(D)$) in order to obtain chaotic behaviour of the system could be somewhat relaxed if the excess saving function $\hat{F}(Y)$ depended on both the level and the rate of change of the representative variable Y. To see this, let us modify the investment function of the Arneodo-Lorenz model by postulating that investment depends not only on the level but also on the rate of increase of income: in other words, by introducing an 'acceleration' effect. The saving function will be left unchanged. For simplicity's sake, we shall assume that the investment function $I(Y, \dot{Y})$ is separable, convex in the first argument and linear in the second, namely:

$$I = I(Y, \dot{Y}) = I_1(Y) + I_2(\dot{Y}),$$

$$I_1'(I) > 0, \quad I_1''(Y) > 0, \quad I_1(0) > 0,$$

$$I_2(\dot{Y}) = v\dot{Y}.$$

where $v > 0$ is the acceleration coefficient, assumed to be constant.

The impact of the 'accelerator' on the dynamics of the system can be readily seen. Including the acceleration effect, our system will now be

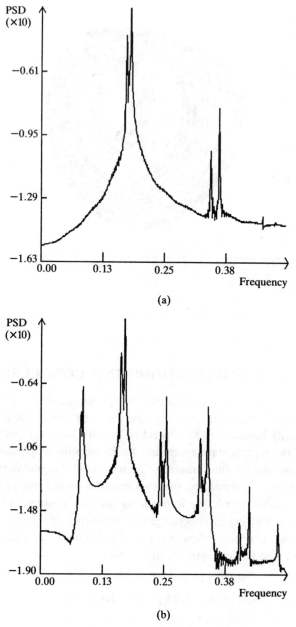

Fig. 13.6. Changing PSD of system (13.5) when the parameter r is varied.

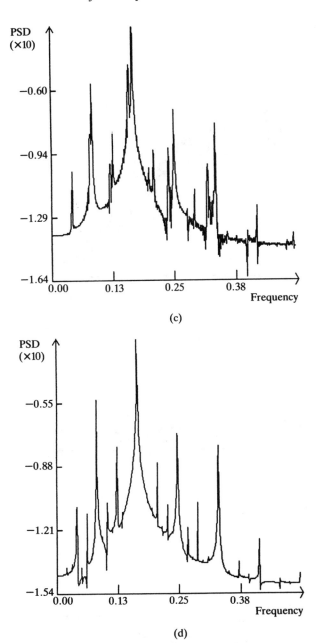

(c)

(d)

Fig. 13.6. (*cont.*)

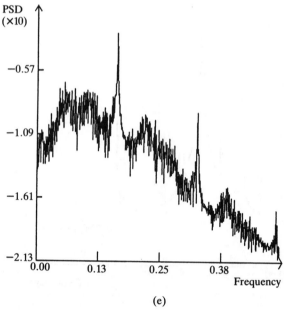

(e)

Fig. 13.6. (*cont.*)

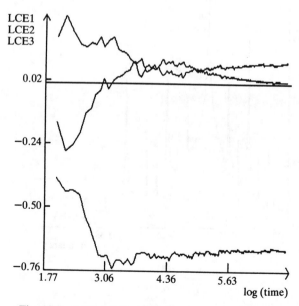

Fig. 13.7. LCEs of the chaotic attractor of system (13.5).

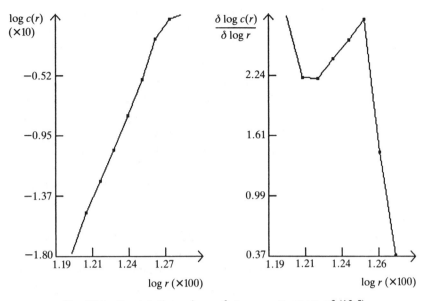

Fig. 13.8. Fractal dimensions of strange attractors of (13.5).

written as

$$G_2(D) = F_1(Y, DY), \qquad (13.9)$$

where $F_1(Y, DY) = \hat{F}(Y) - \hat{v}DY$, and $\hat{v} = (v/ba_2)$, which, in turn, is equivalent to

$$G_3(D) = \hat{F}(Y), \qquad (13.10)$$

where $G_3(D) = G_2(D) + \hat{v}DY$.

Recalling our discussion of the diagram of Fig. 13.2a above, let us now fix $r = \bar{r}$ and choose $c_2 < \bar{c}_2 = (4\bar{r})^{1/3}$, $\hat{v} = 0$ and c_1 small so that we are on the left of the chaotic zone, i.e., in the unstable zone. Clearly, by increasing $\hat{v}$ sufficiently, we can move into the chaotic zone. In fact, for example, by putting $r = 0.8$, $c_2 = 0.4$ and $c_1 < 1$ we can always choose $\hat{v} = 1 - c_1$ so that the exact numerical results of Arneodo *et al.* are reproduced, and chaotic output is generated. Notice, however, that increasing $\hat{v}$ further we shall move into the stable zone. Thus, a large acceleration coefficient may have a stabilizing effect. This rather counterintuitive result is presumably due to the specific lag structure through which the economy responds to inputs.

We conclude that the addition of an acceleration component to the investment function of a macromodel of inventory cycles may modify the dynamic behaviour of the economy substantially. More generally, when the input to a lag mechanism is velocity-dependent, the time profile of the input-output interaction may be altered so as to lead to a more complex motion than would otherwise have been the case. And this even when the additional input in question is linear.

One may also wonder what difference it would make if we used 'complete' Kaldorian saving and investment functions (cf. Fig. 13.1 above).[13] The numerical simulation indicates that in this case the behaviour of the system retains the same broad characteristics, the main difference being that chaotic attractors now remain stable for values of parameters which, with 'reduced' saving–investment functions, would have implied an 'explosion'.

From an economic point of view, the exercises performed in the preceding pages have a mainly pedagogic significance and should not be taken too seriously. However, they give us an interesting clue to the role played by lags and nonlinearities in producing irregular fluctuations and chaos, and confirm the possibility of complex behaviour in simple, low-dimensional continuous-time systems. Firmer results will perhaps be obtained when the analytical and numerical methods discussed so far are combined with a deeper understanding of the mechanism controlling the structures of lags and nonlinearities in real economies. For this purpose economists should perhaps rely more on empirical observations and classifications, and less on *a priori* deduction from very general 'first principles'.

---

[13] Formally this could be achieved, for example, by putting $\bar{Y}_3 = 2$ and replacing the function $\hat{F}(Y)$ with

$$\tilde{F}(Y) = \begin{cases} \hat{F}(Y), & \text{for } 0 \leq Y \leq 1 \\ r(Y-1)(2-Y), & \text{for } 1 < Y \leq 2 \end{cases}$$

14

Analysis of experimental signals –
applications

14.1 DETECTING DETERMINISTIC CHAOS

We shall now turn to the problem of detecting deterministic chaos in
economic time series. As we have already indicated in the Introduction
to this book (Chapter 1), this is a recently developed question on which
economists' opinions differ widely, and no general consensus has yet been
reached. In this chapter, we shall address the issue, making use of the
analytical and numerical methods discussed in Chapter 10. To introduce
our discussion, however, it will perhaps be expedient to provide a concise
and selective overview of the literature.

Economists have tried to apply the techniques concerning chaotic
dynamics essentially to two types of time series: macroeconomic data
and financial data. We shall deal briefly with each of them in turn.

Macroeconomic series
Although from a theoretical point of view, it would be extremely interesting to obtain empirical verification that macroeconomic series have
actually been generated by deterministic chaotic systems, it is fair to
say that those series are not the most suitable ones for the calculation
of chaos indicators. This is for at least two reasons. First of all, the
series are rather short with regard to the calculations to be performed,
since they are usually recorded at best only monthly; secondly, they
have probably been contaminated by a substantial dose of noise (this
is particularly true for aggregate time series like GNP). We should not
be surprised, therefore, that exercises of this kind have not yet led to
particularly encouraging results.

Brock and Sayers (1988) (BS) have studied various US macroeconomic
series (employment, unemployment, industrial production, gross private
domestic investment, GNP), finding some rather strong indications of
nonlinearity, except in the case of GNP, but rather slim evidence of the
presence of chaos in all the series. Brock and Sayers' work provides

estimates of the correlation dimension and of the dominant Lyapunov exponent, as well as new statistics developed by Brock, Dechert and Scheinkmann (1987) (BDS).

The BDS method is intended to test the null hypothesis that a time series is generated by a sequence of i.i.d. random variables against the alternative that the series is generated by a nonlinear deterministic system. The method is based on the concept of correlation dimension suggested by Grassberger and Procaccia (1983). The correlation integral calculated from a time series is defined thus:

$$C(T, m, \epsilon) = \lim_{T \to \infty} \frac{1}{T^2} \sum_i \sum_{j>i} H(\epsilon - \| x_i^m - x_j^m \|) \tag{14.1}$$

where $T = N - m + 1$, N is the length of the series, the $\{x^m\}$ vectors are the m-histories constructed from the time series itself, H is the Heaviside function, and ϵ is a small positive quantity. Brock, Dechert and Scheinkmann (1987) prove the following two results, which hold asymptotically and for every $m > 1$:

$$\sqrt{T}\{C(T, m, \epsilon) - [C(T, 1, \epsilon)]^m\} \xrightarrow{d} N(0, \Sigma), \tag{14.2}$$

$$\sqrt{T}[S(T, m, \epsilon) - C(T, 1, \epsilon)] \xrightarrow{d} N(0, \Omega), \tag{14.3}$$

where $S(t, m, \epsilon) \equiv C(T, m, \epsilon)/C(T, m - 1, \epsilon)$ (the expressions for Σ and Ω can be found in the cited article by BDS). After calculating these statistics, we can divide them by their standard deviations so that they are distributed as normal standardized random variables, and compare the values thus obtained with those of the random variable itself. A precondition of this exercise is that the available data are sufficient, not so much to ensure the convergence in distribution to a normal random variable, as to calculate with sufficient precision the necessary quantities. For this purpose, we must say that with a finite (and often limited) number of data, a correct procedure is possible only when the relevant embedding dimension is rather small, otherwise we run a serious risk of obtaining values of the correlation integral which are smaller than the correct ones. This risk could be avoided by choosing a larger value of ϵ, but then problems would arise for the lower dimensions. On the other hand, it does not seem to be a correct procedure to choose a different scale for each dimension (see also Ramsey and Yuan, 1989; Ramsey *et al.*, 1990).

In this test, the choice of ϵ is particularly important. If we assume that the available data have been contaminated by a noise distributed, say, uniformly on the $[-a, a]$ interval, it then follows that the correlation dimension is equal to the dimension of the embedding for $\epsilon < a$ (i.e., we have the typical behaviour of a purely random process), whereas it is equal to the actual dimension of the attractor for $\epsilon > a$ and $m > 2d + 1$, where d is the dimension of the subspace containing the attractor itself. In order to obtain a reliable estimate of $C(\cdot)$, it is therefore important to choose a value of ϵ greater than the noise level, otherwise the test will necessarily accept the null hypothesis even though the series is chaotic (Scheinkman and LeBaron, 1989a). It is also suggested that different values of ϵ be chosen as fractions of the standard deviation of the available series. If for small values the test accepted the null hypothesis and for greater values refuted it, we could conclude that the series derives from a nonlinear process and has been contaminated by noise.

Brock (1986) suggests an alternative line of action, namely, to adapt to the series under investigation an $AR(p)$ model of the type

$$x_t = \phi_1 x_{t-1} + \phi_2 x_{t-2} + \ldots + \phi_p x_{t-p} + \epsilon_t \tag{14.4}$$

and then to perform all the calculations on the residuals. In particular, Brock proves the following theorem.

Theorem 14.1. *Let $\{x_t\}_{t=1}^{\infty}$ be a sequence of data which are deterministic and chaotic, and let us adapt to it an $AR(p)$ model with finite p; then, in general, the correlation dimension and the dominant Lyapunov exponent of $\{x_t\}$ and $\{\epsilon_t\}$ are equal.*

The proof is a consequence of the invariance principle, according to which if we take a smooth function of a certain vector of data, the topological properties of the series will not be changed. In this case, the residuals are in fact a smooth function of the original series and therefore their correlation dimension and dominant Lyapunov exponent should be equal to those of that series. Brock and Sayers (1988) apply the said method to various economic series, coming to the conclusion that the correlation dimension for the residuals is always much greater than that of the original series and consequently one cannot claim that those series are chaotic. According to the authors, however, these rather discouraging results might be due to the presence of rather strong noise in the data.

Rather similar results have been obtained by Scheinkman and LeBaron (1989b), who studied US series of GNP and industrial production. On the other hand, Frank and Stengos (1988) did not find any evidence of nonlinearity in Canadian macroeconomic series. In another article (1988), the same authors calculated the dominant Lyapunov exponents pertaining to a similar series and always found negative (though very small) values, which apparently indicates that those series are not generated by a chaotic system. As a matter of fact, these estimates raise some doubts: first, because of the very small number of available data (less than 200); and, second, because of the algorithm employed, which is not the classical one of Wolf *et al.* (1985), but a different, rather simplified one directly based on the definition of the Lyapunov exponent. For these reasons, Frank and Stengos' estimates should be regarded with some suspicion and their conclusions cannot be taken as definite.

A particularly interesting exercise has been performed by Barnett and Chen (1986) (BC), who have found results supporting the hypothesis of chaoticity in relation to certain US monetary series. These authors (like Chen, 1988a) used weekly series consisting of the so-called 'Divisia monetary aggregates', which are second-order approximations to the theoretically exact monetary aggregation. (The Laspeyres or Paasche indices are only first-order approximations, and the commonly employed M1, M2 etc. aggregates are not even that.) The Divisia index, both in continuous and in discrete time, calculates the rate of growth of the aggregate as a weighted average of the rates of growth of the components, where the weights are given by the expenditure shares (see Barnett, 1980, 1987). As BC point out, in this case the data are of an unusually good quality as far as economics is concerned, and this favourably affects the estimates of the correlation dimension and the dominant Lyapunov exponents. As concerns correlation dimension, for almost all the series investigated, Barnett and Chen found values of approximately 1.4–1.5 which were strongly convergent. We observe that these values are much lower than those obtained by practically all the other authors, who generally found dimensions between 6 and 7 and a much weaker convergence. It would therefore seem that good quality data is an essential precondition for the effective detection of chaos in economic time series. Alternatively, when this precondition cannot be guaranteed, it then becomes very important to be able to employ filtering methods which are consistent with the requirements of chaos theory. One such method has been discussed in Chapter 10 (section 10.4).

Finally, BC calculated the dominant Lyapunov exponent, which was not usually done in the previously mentioned works owing to the difficulty of computing this quantity in the presence of substantial noise. In this case, too, the results obtained are very interesting, almost always with estimates rather strongly convergent to small but positive values. Once again, the possibility clearly emerges of obtaining satisfactory results and supporting the hypothesis of chaoticity, providing that the data are good to begin with, or that their quality can be improved through filtering.

Financial series

The analysis of financial series has led to results which are as a whole more interesting and more reliable than that of macroeconomic series. This is probably due to the much larger number of data available and their superior quality (measurement in most cases is more precise, at least when we do not have to make recourse to broad aggregations as in the case of share indices).

Scheinkman and LeBaron (1989a) studied US daily and weekly series employing the BDS statistics, and they found rather strong evidence of nonlinearity and some evidence of chaos (the correlation dimension for the weekly data appears to be reasonably convergent to a value between 5 and 6). In order to verify the presence of a nonlinear structure in the data, they also suggested employing the so-called shuffling diagnostic. This procedure consists in studying the residuals obtained by adapting an AR model to a series according to Brock's methodology, and then 'reshuffling' the data. If the residuals are totally random, i.e., if the series under scrutiny is not characterized by deterministic chaos, the dimension of the residuals and that of the shuffled residuals should be approximately equal. On the contrary, if the residuals are chaotic and have some structure, then the reshuffling must reduce or eliminate the structure and consequently increase the correlation dimension. The application of this technique has given good results. Indeed, the correlation dimension of the reshuffled residuals always appeared to be much greater than that of the original residuals. Consequently, it is fair to conclude that, with regard to this kind of data, Scheinkman and LeBaron found evidence not only of nonlinearity but also of chaos.

Some very similar results have been obtained by Frank and Stengos (1989) investigating certain time series concerning the markets of gold and silver. The estimate of the correlation dimension was between 6 and 7 for the original series and much greater and non-converging for the reshuffled data. Altogether, these and similar results seem to suggest that financial series provide a more promising field of research for the methods in question, even though their theoretical relevance may be less than that of macroeconomic series. This is probably due to the greater abundance and quality of data available, which is an absolutely necessary condition in order to calculate the most important indicators of chaos, at least with the algorithms known at present.

14.2 APPLICATIONS OF SSA. NOISELESS DATA

The discussion of the preceding section has further strengthened our conviction that, when dealing with economic data, the presence of noise is a particularly serious problem, and that an effective filtering procedure is therefore an essential step in the attempt to extract correct information from data and to estimate correctly certain classical indicators of chaos, such as LCEs or correlation dimensions. For this purpose, the method we have discussed in Chapter 10, namely SSA, seems to be especially promising and worth testing.

We shall first verify the effectiveness of SSA by applying it to the study of a pseudo-experimental series (i.e., a series generated by a known model). Next we shall extend the trial to true experimental data, derived from certain financial sources.

We shall start with a univariate series derived from the numerical integration of the model discussed in Chapter 11, which, if the integration is sufficiently accurate, can be considered an essentially noiseless set of data. Here we shall make use of two somewhat different types of analysis. The first of these will be applied to the differentiated series and the second one to the detrended series. As we shall see, the two procedures will lead to fairly similar results.

Specifically, we shall investigate a series consisting of 5,000 observations derived from the system of o.d.e.

$$\left(\frac{D}{n} + 1\right)^n x = rx(1 - x), \qquad n = 10, \ r = 5, \qquad (14.5)$$

In particular, we shall consider the series of the variable x_1 obtained by a Runge–Kutta numerical integration of (14.5) with a 0.005 stepsize.

This exercise is mainly aimed at illustrating the differences between the methods of reconstruction of attractors based, respectively, on the Takens method and on SSA. In Fig. 14.1 we show three projections of the attractor reconstructed by the Takens method, with an embedding dimension equal to 7 and a time lag equal to 10. We can see that, as discussed in Chapter 10, the resulting plot is squeezed onto the bisectrix of the positive orthant, and that the first and the third projections, i.e., the projections on the (1,2) and (2,3) planes, are absolutely identical. In order to overcome these drawbacks, we have applied the SSA method, making use of a 7-window obtained by keeping the sampling interval unchanged (i.e., putting $\tau_s = 1$) and choosing an embedding dimension equal to 7. In Fig. 14.2, we can see the diagrams corresponding to the first three columns of the trajectory matrix, projected onto the new

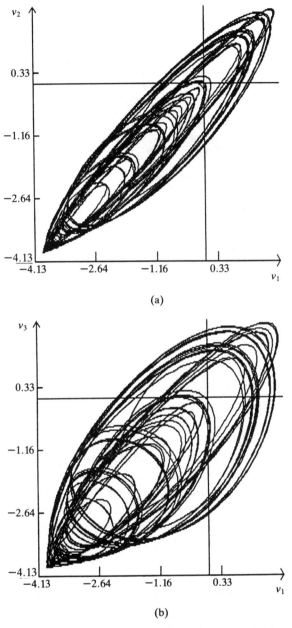

(a)

(b)

Fig. 14.1. Projections of the reconstructed attractor (Takens method).

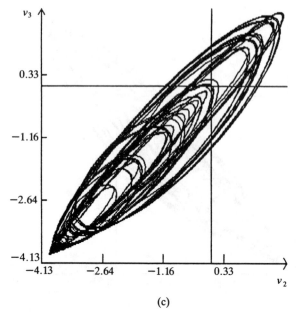

(c)

Fig. 14.1. (*cont.*)

basis. In practice, they correspond to the diagrams of Fig. 14.1. It is immediately obvious that there is no longer any 'squeezing' and that to different planes there correspond different geometrical projections.

As the reader will have noticed, we have not paid much attention to the choice of the embedding dimension and the sampling time. This is due to the fact that, as the series is virtually noiseless, the problem of identifying the significant eigenvalues is essentially non-existent.

To show this and the associated fact that there is no noise floor in this case, we have shown in Fig. 14.3a the spectrum corresponding to the series under investigation. On the ordinate axis, we measure the logarithm of the ratio between the eigenvalues and their sum so as to emphasize the fact that none of them is zero. In Fig. 14.3b, instead, we show a similar spectrum obtained this time with a 20-window and an unchanged sampling interval. In this case, too, it is quite clear that no eigenvalue is zero and that the situation from this point of view has not been changed by increasing the embedding dimension. At any rate, the reconstruction of the attractor can be considered very successful, more so than that obtained by the Takens method. We shall now turn to the more difficult problem of noise filtering.

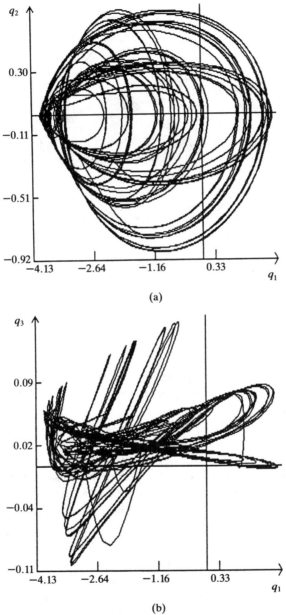

(a)

(b)

Fig. 14.2. Projections of the reconstructed attractor (SSA method).

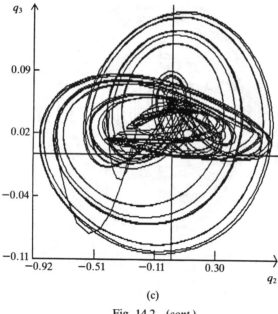

(c)

Fig. 14.2. (*cont.*)

14.3 RECONSTRUCTION OF ATTRACTORS WITH NOISY DATA

In this section, we shall consider the same series as in section 14.2 to which we have added a white noise uniformly distributed over the interval $[-0.3, 0.3]$. Thus, in practice, the signal/noise ratio is approximately equal to 3, i.e., it is rather small. In other words, the noise is rather strong. The consequence of this is immediately seen when we try to reconstruct the attractor by means of the Takens method and project it on the (1,2) plane. From Fig. 14.4 it is obvious that the attempt has failed; the diagram is a tangle which fails to reproduce most of the geometrical properties of the orbits of the system.

Since noise is ubiquitous, the case in question is in some sense more typical than the one considered in section 14.2, which makes the need for an alternative more efficient method even more obvious. For the present purpose, it becomes crucial to choose properly the parameters determining the length of the window, the sample interval and the embedding dimension. First of all, the embedding dimension should be large enough so that the first averaging process discussed in Chapter 10 takes place

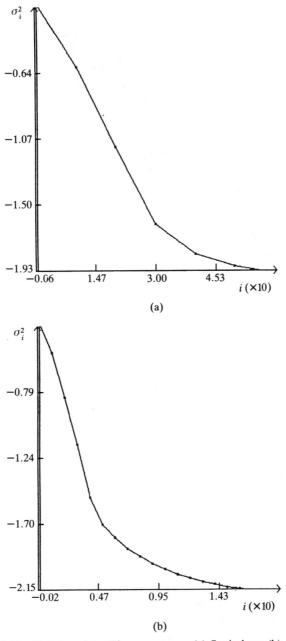

Fig. 14.3. Noiseless data. Eigenspectrum: (a) 7-window; (b) 20-window.

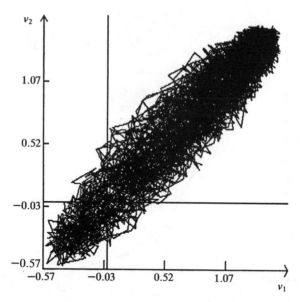

Fig. 14.4. Noisy data. The reconstructed attractor.

over a sufficiently large number of data and also so that the estimate of the covariance matrix may be sufficiently accurate. The window, on the other hand, should not be too long lest high frequency dynamics are obscured. In fact, BK (1986a) propose a method to determine the optimum length of the window; more specifically they suggest fixing it equal to the reciprocal of the minimum significant frequency as indicated by spectral density analysis.

However, to identify such a frequency does not seem to be a simple matter even for series derived from known models, let alone for real series. For this reason, we first calculated the window length by simply identifying the minimum number of points which guarantees some recurrence. In this case, that number is 127, which is too large to cover high frequency dynamics. We then chose a shorter window length, namely 50 observations and we have considered the cases of (i) embedding dimension 50 and sampling time 1 and (ii) embedding dimension 25 and sampling time 2. Fig. 14.5a reproduces the eigenspectrum for the first case, and one can see that there is quite an evident noise floor and that the number of emerging eigenvalues is 7. From this we gather that the statistical dimension associated with the available data is 7, which, not surprisingly, is larger than the dimension of the space where the attractor

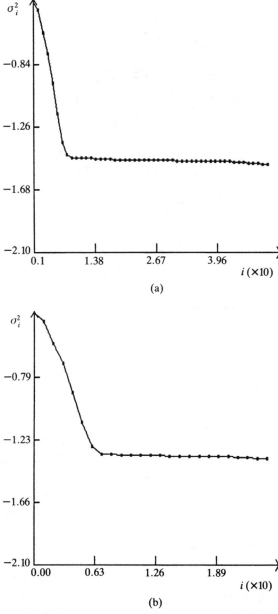

(a)

(b)

Fig. 14.5. Noisy data. Eigenspectrum: (a) embedding dimension 50, sampling time 1; (b) embedding dimension 25, sampling time 2.

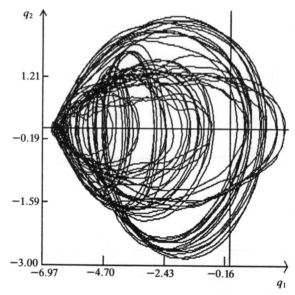

Fig. 14.6. Noisy data. The attractor reconstructed after filtering.

'lives', which, from the analysis of Chapter 11, we already know to be 3. On the basis of the theoretical considerations developed in Chapter 10, we expect the noise floor to change when the sampling interval is altered and, more precisely, that the two quantities should move in opposite directions. From Fig. 14.5b we can see that this is exactly what happens and that, with a sampling interval equal to 2, the noise floor is now higher and the number of emerging eigenvalues is now only 6. By further increasing the sampling interval (with a given window length), we would presumably progressively reduce the number of emerging eigenvalues. But we preferred to filter the original series with the lowest possible sampling interval so that the attractor is reconstructed as accurately as possible.

Fig. 14.6 shows the attractor reconstructed after filtering and projected on the (1,2) plane. It may be noticed that the shape of the attractor is very similar to that of Fig. 14.2a, just a little distorted. This is no wonder considering the substantial amount of noise we have had to eliminate. We can conclude that the filtering method based on the singular spectrum analysis has performed fairly well, even in a very noisy situation. Encouraged by this preliminary result, we shall now turn to the much more challenging problem of applying the method to a real economic time series.

14.4 APPLICATION OF THE SSA ANALYSIS TO FINANCIAL DATA

We shall now consider a series of 4,204 observations of the DM/$ daily exchange rate from January 1973 to December 1989.

When dealing with a real economic time series, the first problem to tackle is the presumably high statistical autocorrelation between observations. The traditional Box–Jenkins approach to this problem is that of differentiating the series one or more times. In the present context, however, we are reluctant to do this because the procedure, besides making the data statistically stationary, is likely to amplify the noise. (This opinion seems to be shared by both Broomhead and King, 1986a, and Vautard and Ghil, 1989.) We have preferred, therefore, to detrend the series, i.e., to regress it against time, so as to eliminate the trend without altering the structure. In any case, we shall compare the results obtained, both differentiating and detrending. As we shall see, the difference between the two results is minimal, which seems to indicate the power of the filtering method adopted. Finally, we have considered the logarithms of the values of the series, as is customary in these cases.

14.4.1 Differentiated series

The first thing to do is to choose the length of the window. This is not a simple matter, but, after a few attempts, we opted for a value identical to that of the preceding section, i.e., 50 observations (corresponding to embedding dimension 50 × time lag 1). Fig. 14.7 shows the resulting eigenspectrum. We can see that the spectrum is almost flat, indicating that noise is apparently very strong in relation to signal. In order to decide which of the eigenvalues are significant, we follow BK's advice and consider the shape of the eigenvectors. In principle, those corresponding to emerging eigenvalues should have a rather regular shape. More precisely, they should have a shape close to that of orthogonal polynomials of degree $i - 1$ where i corresponds to the order of the eigenvector.

Fig. 14.8 illustrates the first eight eigenvectors. We can see that the first seven have a shape which, although somewhat distorted, corresponds in fact to the theoretical one, whereas the eighth and the others that follow (whose diagrams we shall not reproduce) have a rather 'disorderly' shape. This, according to the BK argument, indicates that the eigenvalues

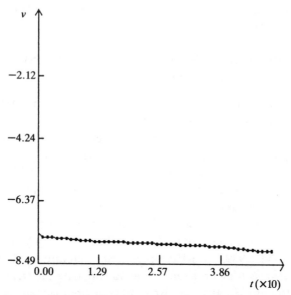

Fig. 14.7. Eigenspectrum of the differentiated series.

corresponding to eigenvectors eight or higher belong to the noise floor and they are not significant. We were therefore led to conclude that the statistical dimension suggested by the data was equal to seven and we therefore filtered the series assuming seven significant directions. The resulting would-be attractor is shown in Fig. 14.9a, whereas Fig. 14.9b shows an attempt to reconstruct using the Takens method. While in the former diagram a certain ordered structure can be recognized, this is not the case for the latter.

Next, we proceeded to evaluate the dominant Lyapunov exponent and the correlation dimension for the reconstructed filtered attractors. Fig. 14.10 shows the Lyapunov exponent apparently converging to 0.12, a value sufficiently large to be significant, especially since convergence looks fairly good. This suggests the presence of a chaotic deterministic component in the DM/$ series. The evaluation of the correlation dimension is an even more delicate task since it is notoriously sensitive to noise, as well as to very long-run dynamics which are possibly present in the series (see, for example, Ramsey and Yuan, 1989). The only information we have on the dimensionality of the generating process is the statistical dimension obtained by means of SSA, i.e., seven.

The estimate of the correlation dimension is ≈ 2.11 and the regression

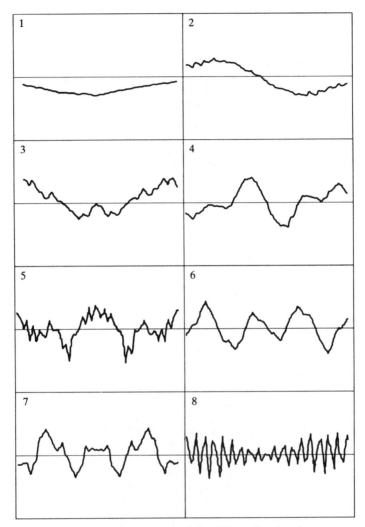

Fig. 14.8. Eigenvectors of the differentiated series.

appears to be fairly good ($\chi^2 \approx 10^{-2}$) (see Fig. 14.11). The result is quite encouraging and again suggests the presence of a low-dimensional chaotic deterministic component of the series. Finally, we thought it would be interesting to examine the power spectrum obtained from the filtered series. From this we did not expect any conclusive evidence that the series was chaotic or random, but rather we wanted to check whether

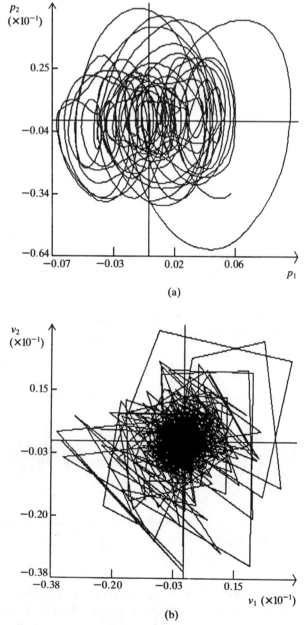

Fig. 14.9. The reconstructed attractor: (a) the SSA method; (b) the Takens method.

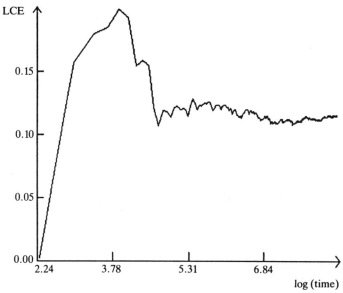

Fig. 14.10. The dominant LCE of the differentiated series.

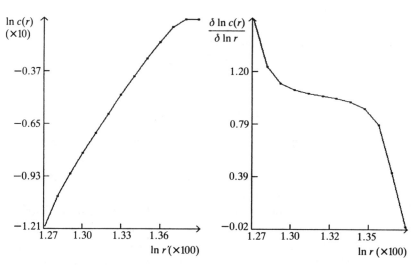

Fig. 14.11. Correlation dimension of the differentiated series.

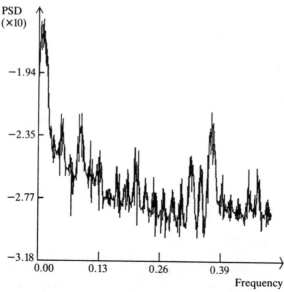

Fig. 14.12. PSD of the differentiated series.

there were strong periodicities in the series. The PSD diagram is shown in Fig. 14.12, which has been constructed from the first principal component of the trajectory matrix X. We can see that there is no indication of either periodicity or low-degree quasiperiodicity. In fact, the spectrum is rather similar to some of those derived from models whose output is known to be chaotic. Together with the other findings, this is also a comforting result.

14.4.2 Detrended series

We shall now investigate our time series, detrending rather than differentiating it. We have, that is, regressed the series on time, so as to eliminate any basic trend which might be present. Fig. 14.13 shows the eigenspectrum constructed using the same parameters as before (i.e., a $50 = 50 \times 1$ window length). It can be seen that there is a fairly distinct noise floor from which eight eigenvalues seem to emerge. To verify this result, we examined the shapes of the eigenvectors, which are shown in Fig. 14.14. They have the approximate shape of polynomials of degree $i - 1$. Next, we filtered the series assuming eight significant directions and projected the result onto the space spanned by the eigenvectors of

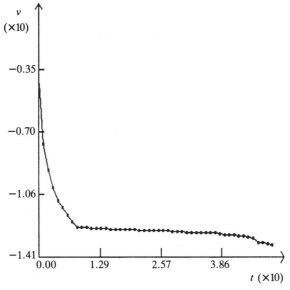

Fig. 14.13. Eigenspectrum of the differentiated series.

the covariance matrix. The result is shown in Fig. 14.15. The resulting would-be attractors look reasonably structured geometrical objects, and in particular the second one, when projected on the (2,3) plane, is fairly similar to that shown in Fig. 14.9a, which is also a comforting result.

14.5 CONCLUSIONS

The analysis of a real economic series has produced rather good results, especially as concerns the reconstruction of the attractor and the calculation of the dominant Lyapunov exponent and the correlation dimension. All these results also agree quite closely with the theoretical considerations developed in Chapter 10. The following general conclusions can be drawn from this investigation:

(i)   Filtering a presumably noisy series before attempting any analysis of the hypothetical attractor is a vital step and for this purpose the SSA provides a very valuable tool.

(ii) In order to obtain valid results, we should analyse sufficiently long series.

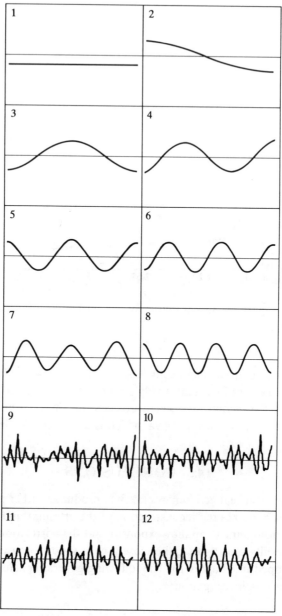

Fig. 14.14. Eigenvectors of the detrended series.

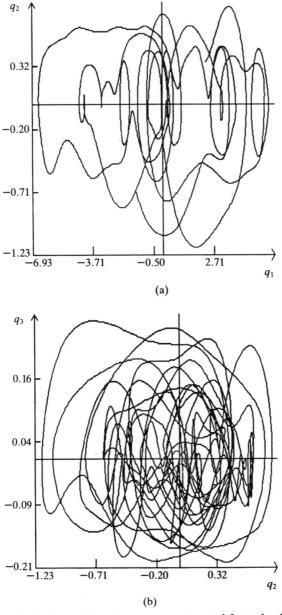

Fig. 14.15. Projections of the attractor reconstructed from the detrended series.

(iii)   We have obtained rather strong evidence of chaoticity in the DM/$
        daily exchange rate.

The optimism induced by these good results should be tempered, however,
by noting three unresolved problems:

(i)   It is still not clear how the filtering method based on SSA affects
        the calculation of correlation dimensions and Lyapunov exponents.
        In particular, the averaging implicit in the projection of the series
        onto a new basis could eliminate certain deterministic components
        of the motion, especially those corresponding to high frequencies,
        thus artificially reducing the estimate of the correlation dimension.
(ii)   It is not clear how reliable the algorithm normally used for the
        calculation of those indicators is, when applied to series which
        are necessarily rather limited and shorter by one or two orders of
        magnitude than those usually employed in the natural sciences.
(iii)   Finally, we notice that the observations of our series were sampled
        daily and we do not know how our results would change if we fur-
        ther reduced the sampling interval. This is a particularly important
        consideration in view of the fact that our analysis of Chapter 10
        shows that the sampling interval is a rather crucial parameter.

Part III

Software

15

DMC Manual

15.1 INSTALLATION

DMC has been designed for two types of fundamentally different personal computers. The first version was produced for IBM-compatible PCs (equipped with the MS-DOS operating system), and the program was then converted for Apple Macintosh PCs.

The difference between the two operating environments is such that there are inevitable, though not fundamental, differences between the two versions. This manual will discuss primarily the MS-DOS version, but will cover both, pointing out any relevant discrepancies.

15.1.1 MS-DOS

A properly functioning DMC requires the MS-DOS operating system version 3.00 or later. The CONFIG.SYS file must, in addition, contain the two instructions:

BUFFERS=20
FILES=20

In this case, one of the following graphics adaptors, is required:

- CGA (Colour Graphics Adaptor). This graphics adaptor allows a maximum resolution (in two-colour working mode) of 640×200 pixels. DMC supports it, even if the necessary resolution requirements make it less than optimal for the type of functions to be performed;
- EGA (Enhanced Graphics Adaptor). In this case the maximum resolution is of 640×350 pixels. DMC does not exploit the colour graphics possibilities of this adaptor as it does with the VGA which we will discuss later. It can therefore function with the various existing versions of EGA (EGA 64Kb, EGA 256Kb, EGA Black and White);
- Hercules. This graphics adaptor allows a resolution of 720×348 pixels;

– ATT/Olivetti/Compaq portable (gas-plasma display). This allows a two-colour maximum resolution of 640×400 pixels;
– VGA (Video Graphics Array). This type of graphics hardware, which is increasingly widespread, allows a resolution equal to 640×480 pixels. It is used in the two-colour mode.

In addition, the minimum working memory required for running DMC is 500Kb. If there is insufficient memory the graphics and computation capacity will be limited. A mathematical co-processor is useful, given the type of computations to be performed.

The installation is performed with the INSTALL program (provided with DMC). The user must first of all insert the DMC diskette in a free drive of the PC, which will be called drive A. The program must then be installed with the following command:

A:\INSTALL

The program will then request the following information:

– the name of the current drive in which the DMC diskette is placed (in this case, A:);
– the name of the subdirectory of the target drive in which DMC is to be installed (we shall call it C:\DMC);
– the type of graphics hardware in use. If this is not known, it can be found in the user's manual provided with the PC;
– the type of printer connected. DMC provides for three types:

    Epson
    Hewlett Packard (HP) Laserjet II
    PostScript

At the end, INSTALL will go on to copy the DMC files onto drive C, in the directory C:\DMC. During this process it will create the following directories:

  – C:\DMC\TMP. This directory is used for memory swapping;
  – C:\DMC\MODELS. This is the directory in which the models created with DMC are stored. Some 'classic' models (such as Hénon, Lorenz, etc.) are also copied to this directory during the installation procedure;
  – C:\DMC\DATA. All the DMC output files are stored in this directory;

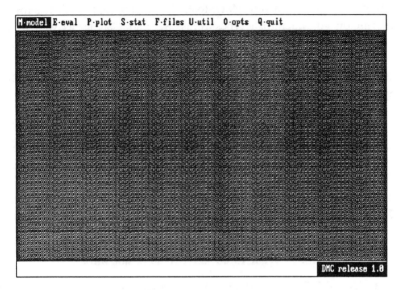

M·model E·eval P·plot S·stat F·files U·util O·opts Q·quit

DMC release 1.8

Fig. 15.1. DMC main menu.

- C:\DMC\EXC. All files 'exported' from DMC to ASCII format are stored in this directory;
- C:\DMC\PLOTS. DMC allows the bit-mapped representation of any produced graphics to be saved in a file. Such files are stored in this directory.

Finally, all information needed for running DMC is stored in the file CONFIG.DMC.

At the end of the installation procedure the user gives the command 'DMC' which will initiate the program and bring the header (copyright notice) onto the screen. By pressing any key, the DMC main menu will be displayed (Fig. 15.1).

15.1.2 Macintosh

The program requires Finder, version 6.1.4 or equivalent, and System 6.0.4 or equivalent. The installation can be performed simply by copying the DMC folder on the program diskette to the root-directory of the hard disk of a Macintosh system. The folder contains the INSTALL program and two versions of DMC, named respectively dmc68000 and dmc68020. The latter requires the presence of a Motorola 68020 CPU, or upgraded versions of it (for Macintosh II, IIx, IIcx etc.), while the former can run on all Macintosh types (Plus, SE and Classic, as well as the models indicated above). Before running the desired version of DMC, use INSTALL to create the CONFIG.DMC file, which will contain the same information about the DMC work directories as in the MS-DOS version. At the startup DMC will show the header (copyright notice) and, after a mouse-click, will lead the user to the main menu.

Notice that, since DMC cannot work with colours, the user must use the Control Panel Monitors command to switch the display mode to black and white. Notice also that the maximum resolution supported by DMC is 640×480 pixels (which is standard on the Macintosh II series). For example, we founded impossible to operate DMC with a Macintosh FX machine endowed with a 16 million colours graphics environment.

15.1.3 What to do if DMC does not work

MS-DOS. DMC's working may be impaired, totally or partially, by the presence of certain special programs (such as 'terminate and stay resident', TSR) or drivers. In this case the only remedy is to remove those programs from the AUTOEXEC.BAT or CONFIG.SYS files.

Occasionally, problems may arise when the SHARE program is in use together with DMC. In this case, the user may try to alter the parameters of SHARE. For example, on one occasion, we solved the problem simply by inserting the command 'SHARE /F:20000/L:50' into the AUTOEXEC.BAT file. Problems may also be caused by the mouse-operating programs. Finally, undesired sound effects may be caused by the BIOS in use.

Macintosh. Analogously, DMC's working may be impaired by the presence of certain INIT programs. The remedy is again their removal.

15.2 GRAPHICAL USER INTERFACE

Before introducing the actual commands, we shall explain the procedures to activate them. In other words, we shall describe how the user interacts with DMC.

DMC is equipped with a user interface designed to simplify input/output operations as well as maximize their speed. It is built on a modern data processing philosophy of communication with the user, and exploits the opportunities for speed, memory and graphics offered by

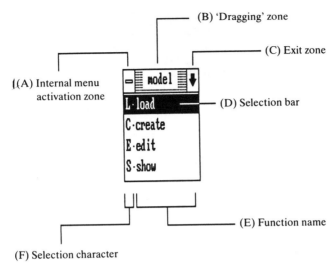

Fig. 15.2. Menu example: the MODEL menu.

modern PCs. The use of the mouse allows faster access to the program's resources, although there is only a small time penalty for keyboard users.

Menus

The functions are listed in entries presented in a menu format. An example of a menu is shown in Fig. 15.2.

The functions illustrated in the figure may be selected in different ways.

- First of all, the user can run the selection bar (D) along the menu (by means of the cursor keys) and then press the <RETURN> key once the bar is positioned on the desired function;
- A more rapid, if less intuitive, access to the functions is achieved by pressing the key corresponding to the selection character (F);
- The left key of the mouse may be pressed after the pointer has been positioned on the desired function (this operation of 'pointing and pressing the key' will be referred to as *selection*).

What follows regarding zones A, B, and C is also valid for other types of 'objects' to be introduced later although this information will not be repeated.

- *Internal menus* (zone A). Many menus employ a zone for activating the 'internal menu'. This term refers to a menu which contains

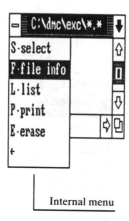

Fig. 15.3. Internal menu example.

functions essential for the particular use of what is contained in it. An example is shown in Fig. 15.3. In addition to selection with the mouse, it is possible to call it by pressing the function key <F2>.

– *Menu shift* (zone B). The menus, like many other items, can be shifted. This can be done in two ways: (a) selecting zone B and moving the menu by keeping the left key pressed. Note that the menu is not immediately moved. What the user is moving on the screen is only a rectangle including the menu area. Only when the key is released does the real movement take place; (b) utilizing the combination of keys <ALT–M> (press the <ALT> key and then, without releasing it, the key M) and moving the menu with the help of the cursor keys. Once the desired position is reached, press the <RETURN> key. Note that only the principal menus maintain the new position after exiting. Zone B is also intended to contain the title of the window.

– *Exit from the menu* (zone C). The user can exit from the menu by either pressing the <ESC> key or selecting the zone C exit with the mouse.

Help

DMC is equipped with an on-line HELP linked to every function of the menu. This allows the user to obtain complete information about the function to be used, and is activated by the following two steps.

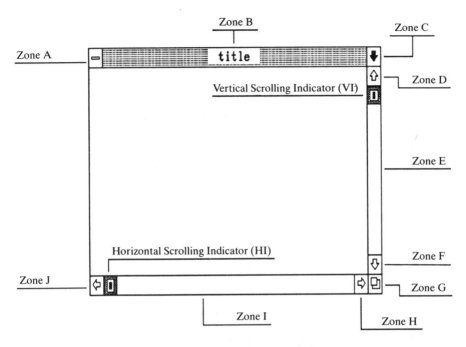

Fig. 15.4. An input/output window.

(1) Bring the selection bar (D) to the desired function
(2) Press the <F1> key.

Input/output windows. Graphic windows

These windows are used both for the input of data (input/output) and for display (graphics). A typical set-up of a window is shown in Fig. 15.4.

Zones A, B, and C are analogous to those discussed above, so only the remaining zones will be considered in detail.

– *Vertical scrolling* (zones D, E and F). Selecting zones D or F with the mouse makes the scrolling indicator (VI) move upwards or downwards, while the contents of the window move in the opposite direction from that indicated by the arrows within it. The user can also make the contents of the window scroll by selecting (VI) directly and thus dragging it in the desired direction without releasing the left key of the mouse. Keyboard users can use the cursor keys (Fig. 15.4).

- *Dimensioning* (zone G). The input windows and graphics which have a zone G can be redimensioned in relation to the space still free on the display. Mouse users can carry out this operation by simply selecting zone G and then 'dragging' it into the desired position. This function is also activated by the combination of keys <ALT-S>. A new dimension for the window can be obtained by using the cursor keys. Once the desired dimension is reached, the <RETURN> key is pressed.
- *Horizontal scrolling* (zones H, I and J). These three zones do the same job, for horizontal movement, that zones D, E and F do for vertical scrolling. In fact, selecting zones J or H with the mouse causes movement of the scrolling indicator (HI) towards the right or left, respectively, while the contents of the windows move in the opposite direction to that indicated by the displayed arrows (Fig. 15.4). The contents of the window can also be moved using the horizontal scrolling indicator (HI) in the same way as described for vertical scrolling.

File selection

Most functions available on DMC require the input of a filename from which to take the data. This can be specified, in addition to writing it directly, by using another file selection procedure offered by DMC. This procedure can be called by the user by pressing the function key <F9>. This selection mode will be called the 'indirect method'. The indirect method displays a directory of working files relevant for the function in use (it is, therefore, 'context-sensitive'). File selection depends on whether the mouse or the keyboard is being used.

- If the mouse is used, selection is performed by simply pressing the left key on the filename desired, and then pressing <ESC>.
- If the keyboard is used, the procedure is: (a) scroll the list of files with the cursor keys until you bring the desired file-name into the upper part of the window (b) press one of the two <SHIFT> keys and (c) press <ESC>.

Macintosh. The function keys <F1> to <F10> are not present on Macintosh (Plus, SE, Classic) keyboards. They are substituted, respectively, by the key combinations <COMMAND–1>, <COMMAND–2>,...,<COMMAND–0> (see the section below for a definition of 'key combination').

15.3 THE EDITOR

A Wordstar-like editor has been provided to allow easy data input and output. Wordstar has been selected as a model because of its widespread popularity as an editor and word-processor.

DMC's editor possesses two working modes. The first, and most common, allows the input of the data necessary for the various available functions and will hereafter be called 'mode 1'. The second, in contrast, allows the input of text without any of the constraints imposed by mode 1. This is used only in the course of the model description and will be called 'mode 2'. The two procedures share most commands and are distinguished only by certain features which underlie the specific task for which they are designed. We first discuss commands which are common to both mode 1 and mode 2.

Before beginning, it would be best to explain that, following a now standard practice, the notation <key> means the pressing of the key indicated between angle brackets. The result of this operation is, generally, a non-ASCII character. The notation <key 1–key 2>, referred to as 'key combination', means pressing and holding key 1 while key 2 is pressed, without releasing key 1. Example: <CTRL–T> means pressing and holding the <CTRL> key followed by pressing the <T> key.

- ASCII characters possessing codes in the range 32–127 may be selected by pressing the corresponding key or the combination <ALT–ASCII code>;
- <↑>, <CTRL–E> moves the cursor upwards from the current line;
- <↓>, <CTRL–X> moves the cursor downwards from the current line;
- <←>, <CTRL–S> moves the cursor one character to the left;
- <→>, <CTRL–D> moves the cursor one character to the right;
- <INS>, <CTRL–V> selects or de-selects the insertion mode. The cursor changes form according to the mode selected: note that the insertion mode moves all characters which are to the right of or atop the cursor to the right as new material is added;
  Macintosh. The <INS> key is the second key (from left to right) of the upper key-row of the numeric keypad.
- <DEL>, <CTRL–G> deletes the character on the cursor;
  Macintosh. The <DEL> key is the upper-left one of the numeric keypad.

– <BACKSPACE>, <CTRL–H> deletes the character immediately to the left of the cursor;

– <HOME>, <CTRL–A> takes the cursor to the beginning of the current line;

– <END>, <CTRL–F> takes the cursor to the end of the current line;

– <CTRL–T> deletes the characters to the right of the cursor and the character underneath it;

– <PGUP>, <CTRL–R> takes the cursor to the first displayed input line;

Macintosh. The <PGUP> key is emulated by '/' of the numeric keypad.

– <PGDN>, <CTRL–C> takes the cursor to the last displayed input line;

Macintosh. The <PGDN> key is emulated by '*' of the numeric keypad.

– <CTRL–K> <B>, <CTRL–K> <K>, <ALT–C> stores the current line in a temporary buffer zone. This information may later be copied on another input line with the following command;

– <CTRL–K> <C>, <ALT–P> copies the input line saved with the previous command onto another line;

– <CTRL–K> <V> moves the stored line to another line;

– <CTRL–Q> <R> moves the cursor to the beginning of the input buffer;

– <CTRL–Q> <C> moves the cursor to the end of the input buffer;

– <ESC> interrupts the input operations. In most cases, it also prompts a request to confirm the input data. This occurs through the internal menu illustrated in Fig. 15.5. By selecting 'confirm' all introduced data are confirmed and the input window is removed. Selecting 'abort' causes the removal of the input window and the abandoning of the DMC function which created its display. Finally, by selecting the last option (←), or equally, by pressing the <ESC> key again, you enter the control mode of the window (the MOVE, DIMENSION, and RUN operations) which is used in the ways described in the section on the use of the user interface. You may return to the input operations by simply pressing the <RETURN> key. With a mouse, you move to the control mode of the window by directly selecting the desired areas described in the section on the user interface.

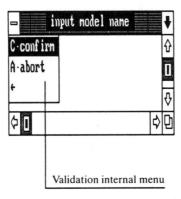

Validation internal menu

Fig. 15.5. The validation internal menu.

Mode 1 commands

– <RETURN> moves the cursor to the next input line. If the current input line is the last, the resulting move is analogous to that achieved by pressing the <ESC> key described above.

In mode 1 an input validation is also performed. If, for example, the user puts a string of alphabetic characters where a number is expected, a warning beep is sounded.

Mode 2 commands

– <RETURN> creates a new input line under the current line and moves the cursor to the line created;

– <CTRL–N> inserts a new line in the place of the current line which is scrolled downwards;

– <CTRL–Y> deletes the current line. It will be replaced by whatever line follows. This generally means that the upper lines of input following the deleted line scroll upwards.

15.4 FUNCTIONS

DMC is structured as a tree of hierarchically arranged menus which begins from the main menu mentioned above. This section discusses commands in the main and in all submenus.

15.4.1 MODEL

The MODEL menu allows access to all the functional capacities which permit quick loading, updating and creation of continuous or discrete dynamic models to be analysed by DMC. It introduces a submenu which contains the following functions:

LOAD

Permits the selection of a model for further processing (EVAL menu).

CREATE

A new model may be created with this function. The user must supply the filename with which it will later be identified. The model is created using the syntax illustrated below.

The user can choose between two basic types of models, namely: continuous-time (differential equations) and discrete-time (difference equations). The choice is performed by means of the instruction 'type', thus:

$$\text{type \{continuous\}}$$

allows the user to apply the procedures for numerical integration of differential equations, while

$$\text{type \{discrete\}}$$

identifies the model as a system of difference equations.

Model parameters are named in the instruction 'parameters'. Example:

$$\text{parameters \{a b c d\}}$$

If no parameters are required by the model, type in

$$\text{parameters \{\}}$$

The variables of the model are introduced by means of the 'variables' instruction. Example:

$$\text{variables \{x y\}}$$

Model equations are entered through the 'model' command. Example:

```
model {
        eq1 = a * x + b * y;
        eq2 = c * x + d * y;
        }
```

Note that the dynamic equations are introduced by the keyword 'eq' followed by a serial number starting from 1.

Each equation is assigned to a variable, keeping to the same order as in the VARIABLES statement. Thus, in the example above we would have:

$$dx/dt = eq1 = a * x + b * y,$$
$$dy/dt = eq2 = c * x + d * y.$$

Model complexity can be reduced using 'intermediate equations'. For example, the model above could be written as

```
model {
    p = a * x; q = b * y;
    eq1 = p + q;
    p = c * x; q = d * y;
    eq2 = p + q;
}
```

As for the syntax of the equations, see the information supplied by the 'syntax help' command listed in the UTIL menu or the details at the end of this chapter.

If the user wishes to evaluate the eigenvalues of the Jacobian matrix or the Lyapunov characteristic exponents, the Jacobian matrix of the model must be given. For example, in the case of the model just discussed, the user would write:

```
jacobian {
    a; b;
    c; d;
}
```

The declaration of the Jacobian matrix is, however, optional.

Comments are introduced between square brackets. Example:

[This is a linear model]

The model, with reference to the examples used up to now, will at the end emerge completely formulated as follows:

```
[This is a linear model]
type        {continuous}
parameters {a b c d}
variables   {x y}
model       {
    eq1 = a * x + b * y;
    eq2 = c * x + d * y;
}
```

```
jacobian    {
             a; b;
             c; d;
             }
```

Notice that, in a model formulation:

(1) the statements can be listed in whatever order the user prefers;
(2) the user can employ upper-case and lower-case letters without con-
     straints.

EDIT

A previously created model may be altered with this function.

SHOW

This function is useful if the user wishes to examine the structure of the
model without altering it.

15.4.2 EVAL

This command introduces a menu containing all the appropriate functions for examining the behaviour of the dynamic model selected. It
allows the user to specify the values of parameters, initial conditions, and
integration procedures. This command also includes other facilities such
as the computation of eigenvalues, Lyapunov characteristic exponents
and bifurcation diagrams. The menu comprises:

PARAMETERS

The values of the model parameters are chosen with this function.

START POINT

The user chooses the initial values of the phase variables, from which the
integration procedure will start.

PREVIEW

During the course of the integration, the user can obtain a preliminary
(two-dimensional) plot of the trajectory being produced. To do this, the
desired variables, as well as the area believed to hold the relevant part
of the orbit, must be specified by means of a pair of co-ordinates.

Example: Suppose that the following model (the LVG model) is to be
evaluated:

$$dx/dt = eq1 = (a - b * y) * x,$$
$$dy/dt = eq2 = (- c + d * x) * y.$$

The values of the parameters are a=2, b=4, c=1 and d=3 and the start point is (x,y)=(1,1). To see the evolution of the trajectory of this system during its computation, one must specify the preview variables

var1   x
var2   y

The minimum and maximum values of the preview co-ordinates must be specified in such a way that computed orbits do not move outside the screen. This of course requires that the user makes a guess on the dynamics of the system. In the present case, one could for example choose:

min var	1	0
max var	1	2
min var	2	0
max var	2	2

INTEG. DATA

This function allows the user to establish the parameters for the chosen integration mode. In the case of a discrete model, they are simply the initial time (t0), the transient part of the trajectory and the number of iterations to perform. The integration parameters are more numerous in the case of continuous models. They are, for the two available algorithms:

- Runge–Kutta:
- t0: initial time;
- stepsize: fixed time interval between two successive evaluations of the trajectory;
- sampling interval: this parameter should only be used to produce data for the PSD algorithm (see the PSD function of the STAT menu), otherwise it should be left blank. When fixing a sampling interval, the user instructs DMC to select a subset of the computed values, i.e., values will be chosen at time intervals equal to (more exactly, not smaller than) the sampling interval. Of course, if the latter is smaller than the stepsize (e.g., zero), all computed points will be chosen;
- transients: transient part of the trajectory (this is not stored in the output file);
- iterations: number of iterations of the given stepsize to perform. In

this case, given the constancy of the stepsize, it is easy to compute the time at which the integration terminates (stepsize × iterations).

- ODEINT:
- t0: initial time;
- tolerance: ODEINT is an integration method with variable stepsize related to the precision of integration desired by the user. The stepsize is then decreased or increased if the integration precision desired is greater or smaller, respectively, than this parameter. A tolerance parameter of approximately 10^{-7} ($1.0e^{-7}$) has in most cases given good results.
- starting stepsize: this is the initial stepsize. It is varied from time to time depending on whether or not the desired precision has been reached;
- minimum stepsize: this parameter must be a very small positive number. It specifies the smallest value of the stepsize to be used in trying to increase the integration precision. The minimum stepsize (m.s.) is closely related to the tolerance parameter (t.p.) discussed above and must be changed proportionally to it. If we choose a very small t.p. and a relatively large m.s., the procedure might stop as it becomes impossible for ODEINT to reduce the stepsize to a value compatible with that of the t.p.
- sampling interval: the options mentioned above allow ODEINT to compute many points in a small interval if a small stepsize is chosen. In order to avoid such a concentration, a sampling interval must be specified from which not more than one point will be taken;
- transients: transient part of the trajectory (this is not saved in the output file);
- iterations: number of desired iterations in the integration process. In this case, one cannot establish the end-point of the trajectory.

GO

The integration procedure is started with this command. The first step is to indicate the file in which DMC will store the data of the computed trajectory. It will be created in the DATA work directory, and its contents can be 'exported' in ASCII (and therefore readable) format with the EXPORT function available in the FILES menu.

At the end of every integration, a menu containing the following commands is activated:

– STOP: interrupts the integration procedure of the model relative to the trajectory undergoing processing;

– CONTINUE: if, for example, some interesting characteristic is observed in the trajectory just obtained, the user may decide to continue the processing, restarting from the last point obtained. The integration parameters specified with the INTEG. DATA function from the EVAL menu will still be valid;

– PROCEED: this function may be used after having specified, with the VARIATION command from the EVAL menu (refer to this for further information), the amount to be added to or subtracted from each variable or parameter of the model in order to obtain other trajectories on the same plot.

Notice that the only way to obtain multiple trajectories on the same plot is to use the VARIATION/PROCEED command combination, after specification of a suitable PREVIEW box.

VARIATION

This function is useful if the user wants to explore the behaviour arising from changes in the values of a system of one or more variables or parameters. It is intended to be used in conjunction with the PROCEED command (available at the end of the integration of each segment of the trajectory) which carries out the indicated variation. VARIATION introduces a submenu by means of which the user can:

(a) activate or de-activate the application of the variation to the integration data;

(b) fix the size of the variation.

Example: Suppose the model (call it TEST) that we are evaluating has these variables and parameters: v1, v2, p1 and p2. If we want to change their values automatically we should input, for example:

variation file prefix	test
p1	0.1
p2	0
v1	0.4
v2	−0.3

Then we should activate the VARIATION procedure (a) and use the GO command of the EVAL menu. This generates a data file named *test.1*. The values of variables contained in *test.1* have been computed for values of the parameters (and initial conditions) which are equal to those originally provided by the user, plus or minus the variations indicated above. To iterate the procedure, we can use the PROCEED command at the end of each computation. This will generate data files called *test.2, 3, ...,* each time modifying the parameters (and initial conditions) as indicated.

LYAPUNOV

The Lyapunov characteristic exponents of the selected model are computed. This is possible only if the Jacobian matrix has been specified. The necessary data are:

– number of time units to be used in sampling each Lyapunov exponent. Each exponent is, in fact, evaluated at every instant and it is therefore often necessary, for reasons of computing efficiency, to take from the total only a limited number of sampled data;
– name of the output file. The variables will be placed in the following order in this file: TIME, LCEi, $i = 1, 2, ..., n$ where n is the dimension of the phase space of the system.

EIGENVALUES

If the Jacobian matrix has been specified, it is possible to compute its eigenvalues. The data specified in the PARAMETERS and START POINT section of the EVAL menu are used for the computation.

BIFURCATIONS

By using this function the user can obtain the bifurcation diagram of the chosen model. In order to proceed, the parameters, the initial state and the integration data must be designated. Once this is done the following must be supplied:

– the name of the parameter to be varied;
– the lower limit of the variation interval of the parameter;
– the upper limit of the same interval;
– the number of subdivisions in the interval. Since the computation of the diagram is very time-consuming, using a small number of subdivisions (200-300) lets the user know if the correct parameters have been selected. To obtain a complete diagram it is suggested that between 450 and 550 subdivisions are used, depending on the horizontal resolution of the graphics environment in use;
– the variable against which to plot the parameter;
– the lower limit of the variation of the variable;

– the upper limit of the same interval.

In the case of continuous models the evaluation of the bifurcation diagram is performed by calculating the variable maxima with respect to time. The relevance of each maximum encountered by the program is established using a further parameter, i.e.,

– a cut-off value below which the maximum is not relevant to the computation.

STATUS

By selecting this function the user is notified as to the current state of the EVAL menu parameters. The information given is as follows:

– name of the selected model;
– value of its parameters;
– initial state of the system;
– the 'preview box' if specified;
– the integration procedure;
– the state of the variation mode.

INTEG. MODE

If the model is of a continuous type the user may choose between two different integration algorithms:

– Runge–Kutta: this method is used by default. It allows the greatest speed though possibly at the cost of lower precision.
– ODEINT: with this algorithm a higher precision of trajectory computation is obtained at the price of a great deal of machine time.

15.4.3 PLOT

The data of any DMC-produced file can be portrayed in two- or three-dimensional plots. The commands available are as follows:

WORK FILE

This function allows the user to specify the file whose data will be used for the graphics available in the PLOT menu.

VARIABLES

The variables to be graphically represented are indicated with this command. For example, to produce a two-dimensional plot starting from a file containing x, y and time variables, the user should write in the input box:

$$x \; X$$
$$y \; Y$$

DMC can perform transformations of the variables contained in the internally generated or 'imported' files. For example, one could write:

$$x \; X\text{^}2$$
$$y \; \log (Y)$$

Notice that, in this example, a quadratic (non-monotonic) transformation has been performed. This could cause the exclusion of some parts of the plot. To avoid this inconvenience the user should specify the interval of values to be plotted, for example:

$$x \; X\text{^}2[0,10]$$
$$y \; \log (Y)$$

where the limits of the plot for the variable along the *x*-axis are included between square brackets. The same applies to transformations of the variables measured on the other axes. In the case of *two-dimensional* transformations it is also possible to represent more than one variable on the ordinate axis, for example:

$$x \; \text{TIME} \; [0, 100]$$
$$y \; X; Y; X + Y; EXP(X) \; [0, 10]$$

Particularly complex transformations may be simplified with the use of intermediate steps. In addition to the two available input lines, others may be added with the EQ.NUM. command from this same menu. This allows the user to increase the number of available lines up to a maximum of five. The user may write, for example:

$$t1 = \exp (X\text{^}2)$$
$$t2 = \sin (Y - X)$$
$$x \; \text{TIME}$$
$$y \; X + 2 * t1 + (1 - t2) + t1 * t2$$

The axes are labelled according to the indications given by the user on the last *n* lines of the input window (*n*=2 for two-dimensional plots; *n*=3 for three-dimensional ones). A different output can be obtained using the following syntax:

$$x \; \text{TIME} \backslash \text{time-interval}[0,100]$$
$$y \; X;Y;X+Y;EXP(X) \backslash \text{misc}[0,10]$$

The *x*-axis will be labelled with the name 'time-interval' while the *y*-axis will be named 'misc'. Notice the use of the backslash symbol (\) and that the re-scaling (implied by the numbers in square brackets) must be indicated *after* the labels.

The names of the variables or the functions displayed in these examples are *not* 'case-sensitive' and the differentiation has been used simply for the sake of clarity. These notes apply, with the already noted exceptions, to three-dimensional plots as well.

DRAW

Uses the data specified by the VARIABLES command, and applies it to the selected WORK FILE to produce two- or three-dimensional plots.

EQ.NUM

The EQ.NUM command allows the user to obtain further input lines in addition to those already available with the VARIABLES command seen above.

LIMITS

Normally, all the available data are represented on the plot. If instead a partial plot is required, the LIMITS function may be used. The first and last data points to be plotted must then be indicated.

PLOT TYPE

This introduces a submenu which allows a selection of the type of graphic presentation to be used for the next 'plot'. There are four possibilities:

- LINE connects successive points with straight lines;
- MARK connects successive points with lines and marks the points;
- POINT plots isolated points;
- BAR produces a histogram representation, e.g., for the auto-correlation function.

There is, in addition, the possibility of:

- suppressing or displaying the co-ordinate axes in the plot;
- correcting the visualization of circles via the 'aspect' command;
- normalizing the data when represented in a three-dimensional plot. This function can be used when there are great differences in magnitude among the variables to plot.

STATUS

This presents the current parameter situation in the PLOT menu. The user is told whether:

– WORK FILE has been selected and, if so, the name of the file is shown;
– the variables to be represented have been designated;
– the axes have been activated or not;
– the aspect correction has been activated or not;
– the three-dimensional normalization has been activated or not;
– limits have been placed by the user on the first and last data point to be plotted (LIMITS function);

The user is also informed about:
– the type of representation chosen (two- or three-dimensional);
– the preselected graphics representation procedure;
– the number of available equations besides those referring to the axes (see the EQ.NUM command).

3D VIEWPOINT

This introduces a submenu with which the user can:
– move from 2D representation to 3D and vice versa;
– choose the point of view for 3D representation. The user can modify the default 3D viewpoint by choosing the function SET of this submenu, which will prompt a 3D 'reference cube'. Then the user can activate an internal menu by pressing <F2> (<COMMAND–2> on Macintosh PCs) or by directing the mouse to the proper icon. For further details on DMC's user interface see GUI HELP in the UTIL menu.

In three-dimensional plots, the axes pass through the centre of gravity of the subspace (a parallelepiped) that contains the plot. Negative co-ordinates will be evidenced by broken lines; positive ones by normal lines. This command displays a three-dimensional reference system on which the following operations can be performed:

X, Y, Z

These allow the rotation of the reference system around the X, Y, Z axes, respectively, or the movement of the observer along them.

RESET

Takes the reference system back to the default position.

+/–

Reverses the rotation or movement along the axes.

MOV/ROT

Switches between rotation and movement of the reference system.

SPEED UP

Increases the speed of rotation/movement of the reference system.

SPEED DN

Decreases the speed of rotation/movement of the reference system.

STATUS

Shows the state of commands of this menu. The commands performing the updating are: MOV/ROT, +/–, RESET.

15.4.4 STAT

DMC is equipped with a group of functions which allow a statistical analysis of data obtained from dynamic models. The following functions are available:

CORRELATION

This computes the auto-correlation of the data from the file specified. The following must be specified:

– name of the data file;
– name of the variable for which the auto-correlation will be computed;
– the first value of the variable (first in the series of the sampled data);
– last value of the variable;
– whether or not the mean must be subtracted from the data;
– whether or not the variable must be differentiated (first order);
– the number of lags to be considered;
– the name of the file which will contain the final result.

The estimator of the auto-correlation used is:

$$\rho(k) = \frac{\gamma(k)}{\gamma(0)},$$

where

$$\gamma(k) = \frac{1}{N} \sum_{t=1}^{N-k} (Z(t)Z(t+k))$$

and k is the number of lags, $Z(t)$ is the variable and N is the number of sampled data for the variable concerned.

PSD-FFT

The computation of the spectral power density for the specified data is performed. Refer to the main text (Chapter 5, section 5.2) for any further clarification of PSD theory. The following must be specified:

- the name of the input data file;
- the variable to be analysed (even if the file is multivariate, the analysis may be performed on only one variable at a time). The names of the variables contained in the files may be obtained with the command FILE INFO;
- the rate at which the variable is sampled;
- the number of estimates of the PSD to compute;
- the variance-reducing factor;
- whether or not the mean must be subtracted from the data;
- whether or not the data must be differentiated (first order);
- the type of lag window (filtering) to be used. It is important to remember that the lag windows are respectively:

(1) Parzen window;
(2) square window (no filtering);
(3) Welch window.

The Parzen window is the most commonly used. Consult the main text (Chapter 5, section 5.5) for further information.

- whether or not the user wishes to perform 'overlapping' (see section 5.5);
- the file in which the computed PSD data will be placed.

Note that the amount of data necessary for the PSD computation, if overlapping is *not* performed, is equal to

4 * (variance-reducing factor) * (number of PSD samples)

otherwise it is equal to

2 * [(variance-reducing factor) + 1] * (number of PSD samples)

DIM-PHASE

This function computes the logarithms of the correlation integrals for data obtained from the numerical integration of a model. The user must supply:

- the name of the input data file;
- the first data point;
- the last data point (putting zero if the user wants to consider all the points that follow in the data file);

– the number of elements of a random subset of the set of data points selected. This subset allows the user to extract a representative sample of the data set under investigation, thus speeding up the computation with a limited loss of information. To consider all the data points, set this number to zero.

– The name of the output data file.

This function generates a bivariate data file. The variables are:

– R = the logarithm of the radius of the sphere related to the correlation integral;

– CR = the logarithm of the correlation integral related to R.

Note that the logarithms are taken to base 2.

DIM-REB

This function computes the correlation integrals for reconstructed attractors. The user must supply:

– the name of the input data file;

– the name of the variable to analyse;

– the first value of the selected variable and ...

– ... the last value (with reference to the data file);

– the minimum value of the embedding (emb_min) and ...

– ... its maximum value (emb_max);

– the name of the output data file.

This function generates a multivariate data file. The variables are:

– R = the logarithm of the radius of the sphere related to the correlation integral;

– CRi (i = emb_min,...,emb_max) = the logarithm of the correlation integral related to R.

Note that logarithms are taken to base 2.

DIM-CALC

This function estimates the values of the correlation dimension. This is done interactively. The program displays two plots. The one on the left shows the data obtained from one of the two related functions: DIM-PHASE or DIM-REB. The plot on the right shows the derivative of that function.

The user can select the data for estimating the dimension in an interactive way, using the plot on the right. This selection can be performed as follows:

• Using the mouse. DMC shows a rectangle which can be 'dragged' by the mouse movement. The rectangle dimensions can be modified using the <+> (for stretching) and the <−> (for shrinking) keys.

• Using the keyboard. The rectangle can be moved as follows:
   cursor keys move the rectangle to the desired position;
   <+> and <−> change the rectangle's dimension as indicated above;
   <PGUP> increases the rectangle's speed of motion;
   <PGDN> decreases the rectangle's speed of motion;
   <RETURN> sets the rectangle position and starts the estimation procedure.

The program performs a linear regression with the selected data. The slope of the regression curve is the estimate of the correlation dimension. In the case of data obtained from the DIM-REB function, the program plots the correlation dimension as a function of the embedding dimension.

By pressing <ESC> the user can interrupt the estimation procedure.

LCE-FET

This function performs the computation of the dominant first-order Lyapunov exponent for a reconstructed attractor. The procedure follows Wolf *et al.* (1985) mentioned in the text (see Chapter 10 above).

The set of inputs for this procedure is the following:

– the name of the input data file;
– the name of the variable on whose data the evaluation is to be performed;
– the number of data points to consider;
– the expected dimension of the attractor;
– evolution time: this is the number of data points to be followed before the program looks for a new test trajectory;
– dt: this parameter is used to normalize the dominant LCE. When performing the calculations on data from a given model, this is the same as the stepsize. When using 'imported data', dt must be set equal to the ratio between the observation interval and the unit of measure of time selected by the user;
– the maximum acceptable separation between test and fiduciary trajectories, beyond which the test trajectory must be replaced;
– the minimum acceptable separation between test trajectory and fiduciary trajectories, below which the discrepancy is believed to be due only to noise;
– the name of the data file that will contain the output of the procedure.

REBUILD

This function performs the reconstruction of an attractor starting from a univariate numerical series. The method is based on Takens's theorem (see Chapter 10 above). It is employed in the computation of both the correlation dimension (DIM-REB) and the dominant first-order Lyapunov exponent. See the text for further information.

The requested input data are:
- the name of the input data file;
- the name of the variable on whose data the evaluation is to be performed;
- the number of data points to consider;
- the guessed dimension of the phase–space where the attractor 'lives';
- the reconstruction time delay;
- dt: the time interval between data samples;
- the name of the data file that will contain the output of the procedure.

POINCARÉ MAP

This function introduces a submenu which allows the computation of the Poincaré map. The commands in the submenu are:

SET VARS

This function allows the user to specify the work file for the computation of the Poincaré map. In addition to this the user must also specify the variables on whose data the computation will be performed and the name of the file in which the output data will be stored.

SET VIEW

This function establishes the viewpoint (for the three-dimensional reference system) from which the user observes the evolution of the variables specified with the SET VARS command, as well as the details of the computation of the Poincaré map. For positioning the reference system, see the 3D VIEWPOINT command discussed above.

SET PLANE

This function allows the user to fix, in a completely intuitive way, the cutting plane on the basis of which the Poincaré map will be computed. To do this, the plane will be graphically represented in the reference system previously oriented with the SET VIEW command. It may be moved with the commands in the relevant internal menu. They are:

MOVE X, MOVE Y, MOVE Z

Move the plane along the X,Y and Z axes, respectively.

ROTATE Y, ROTATE Z

Rotate the plane around the Y and Z axes, respectively.

RESET

Brings the plane back to the default position.

SPEED UP

Increases the plane's speed of rotation/movement.

SPEED DN

Decreases the plane's speed of rotation/movement.

EQUATION

Displays the equations corresponding to the position of the plane.

+/−

Reverses the direction of rotation or movement of the plane.

MAP TYPE

Displays a submenu which allows the choice of the type of Poincaré map to be constructed. The alternatives are:
- RIGHT: Right Poincaré map. The points to be considered are those determined by orbits crossing the surface of section from the right.
- LEFT: Left Poincaré map. The points to be considered are those determined by orbits crossing the surface of section from the left.
- BOTH (RIGHT + LEFT): Poincaré map from right and left. The points to be considered are those determined by orbits crossing from both the right and the left. This alternative is used only as a preliminary step before deciding which crossing side is best for a proper analysis. BOTH is default.

GO

Once all the parameters are set, this command starts the computation of the Poincaré map. Consult the main text (Chapter 4) for further information regarding the theory of Poincaré maps.

1D MAP

This function performs the calculation of a one-dimensional map starting from the data supplied by a previously calculated Poincaré map.

The information to supply in this case is:

- the name of a data file previously obtained using the Poincaré map procedure;
- the degree of the polynomial which DMC will fit to the data supplied in the data file;
- the name of the data file where the data obtained from the 1D map algorithm is to be stored.

15.4.5 FILES

This command leads to a submenu which offers a series of functions designed for exchanging data and, in general, for managing files. They are as follows:

SHELL

This command allows the user to exit temporarily from DMC in order to use other programs within the limits of the available memory. To return to DMC, type EXIT at the DOS prompt and then press any key.

DIR

Introduces a submenu with which the user can examine the contents of the DMC work directories. The commands for this submenu are:

EXCHANGE

Provides a directory of all the (ASCII) data files exported by means of the EXPORT function in the FILES menu. This function is also available (with some restrictions) in the internal menus of the graphics windows.

DATA

Displays the DMC work directory. This directory contains the output files created by the various DMC functions, as well as the files imported by the IMPORT function in the FILES menu.

PLOTS

Displays the directory in which DMC puts the files containing the plots saved by the user. The DMP extension is automatically set for these files.

OTHER

If the user is interested in the contents of directories other than work directories, s/he may display them by using this function. The input required in this case is the *complete* pathname of the file or files desired. For example,

C:\DMC*.*

displays all the files contained in the DMC directory loaded in drive C.

Macintosh. This function is not supported.

MODELS

Displays the directory containing the files created with the CREATE command of the MODEL menu. All the other commands of the MODEL menu refer to this directory.

AVAIL

Displays the amount of central (RAM) memory available and the space available on the disk in use.

LOAD PLOT

The plots obtained by DMC can be saved on disk by the SAVE command in the internal menu, which can be called by pressing <F2>. Saved plots may be displayed and printed by this function.

To print a plot, move to the internal menu of the plot by pressing <F2> or using the mouse, and then use the command PRINT.

BIND

This function allows two (ASCII) files to be combined into one. It is useful, for example, when there are two time series in different files, and the user wishes to create a single 'bivariate' file. In this case, 'h' is used as 'bind mode'. If instead the user wishes to lengthen a series adding the data taken from another series, 'v' is used as bind mode. In both cases the data contained in the first file will be placed first in the output file.

The input data files can be anywhere on the user's (hard or floppy) disks. To reach them the user can use pathnames. However, notice that to indicate a subdirectory one should use backslashes (\) for DOS machines and colons (:) for Macintosh ones. Notice also that, when using Macintosh machines, one must give the entire pathname to a file (including the name of the root directory, i.e., the name of the hard disk) and that the floppy disk drive is not available for such operations.

EXPORT

The files produced by DMC in binary format can be exported in ASCII format using this function. This requires the input of:

- the name of the source file;
- the name of the first variable to be exported;
- the name of the last variable to be exported. This information may be obtained with the FILE INFO command;
- the first co-ordinate and ...
- ... the last to export for the specified variables;
- the name of the destination ASCII file;
- any comments.

IMPORT

An ASCII data file may be used by DMC if it is imported using this function. It requires:

- the name of the input file;
- the stepsize of any 'index' variable. If no index variable is to be used, press <RETURN> or 0;
- the number of series contained in the file. The user must give to each series a name which can be recognized by DMC. The name must be a string of no more than 10 alphanumeric characters, the first of which must be alphabetic;
- the name of the DMC data file for imported data. Note that all data encountered which are not transformable into numbers are passed over without producing error messages.

15.4.6 UTIL

This gives access to functions which themselves may be useful while processing with DMC. They are:

CALCULATOR

DMC has an internal calculator which allows the user to perform simple arithmetic operations. These must be specified on the input line. Calculation results are displayed on the lower left of the screen. See the

SYNTAX HELP command for the syntax and available functions. This function may be called at any time with the <F10> key.

Example: We type on the input line of the calculator window the following data:

10+log(2).

To obtain the result (which will be displayed on the lower left corner of the screen) we must press <RETURN> twice. To perform another operation, we simply press <RETURN> and proceed as before. To abandon the calculator window we press <ESC> and then choose the ABORT command from the internal menu.

FIND ROOT

This option introduces a submenu which allows a choice between two different methods for the computation of the zeros of a function:

- Cardano: finds the roots of third degree polynomials;
- Brent: a more general method than the previous one, Brent finds the zeros of any function once the search interval and the desired accuracy have been specified. The function must be monotonic within the specified interval.

DATA GENERATOR

This function allows a file to be generated containing a sample from a univariate function; the interval (or subinterval) and the number of data points to be sampled must be specified. This is useful for plotting univariate functions. If the function is not defined for a value within the interval, the corresponding value is set at zero and an error message is produced.

GUI HELP

This command provides detailed information on the functioning of the user interface employed by DMC.

SYNTAX HELP

This is a guide to the proper formulation of algebraic equations which are available with DMC. All functions and available operators are listed in the order of evaluation followed by DMC.

EDITOR HELP

This is a guide to the use of the editor for editing models and input data.

15.4.7 OPTS

This menu allows some system parameters to be set, such as the mouse's movement speed and the type of printer. In addition, the mode of data representation can be altered, which can also be done in the PLOT menu. The commands on this menu are as follows:

MOUSE

Prompts the mouse control menu. Users equipped with a mouse can regulate its movement speed according to the graphics hardware available, or to their own needs. The configuration data are stored in the DMC-MOUSE.CFG file in the program directory. The available commands are as follows:

SPEED DN

This command reduces the speed of the mouse movement.

SPEED UP

This command increases the speed of the mouse movement.

RESET

The mouse is brought back to default state.

SOUND

If an error is made, or when an operation is completed, DMC gives off, by default, a brief sound. The user may avoid this by using this function which, alternately, suppresses and activates the sound.

PRINTER

The default type of printer is established at the time of installation. However, it may be altered by means of this command. DMC provides for three types of printer, namely:

- EPSON: this is the most commonly used dot matrix printer.
- HP LASERJET II: indicates the class to which the Hewlett Packard Laserjet II belongs;
- POSTSCRIPT: indicates the class of printers which use PostScript as a control language. The most common amongst these is the Apple Laserwriter.

It is important to note that if the printer is connected to the COM1

serial port the user must issue the DOS commands:
$$\text{MODE COM1:96,N,1,8,P}$$
$$\text{MODE LPT1} = \text{COM1}$$
before booting DMC.

   Note that the parameters given in the first of the two commands above, which set up the data transmission characteristics, refer to the Apple Laserwriter and that they should be changed depending on the type of serial printer in use.

PLOT TYPE

The user is referred to the identical command in the PLOT menu.

Macintosh. The commands contained in this menu are not supported by the Macintosh version of DMC (and, therefore, OPTS is not present in its main menu). The exception is the SOUND command, which is embedded in the UTIL menu.

15.4.8 QUIT

Introduces a submenu which allows a 'controlled' exit from DMC. The submenu provides the options:

- YES: to abandon the program;
- NO: to continue.

15.4.9 Internal menu commands

In addition to the functions discussed so far, there are others grouped in appropriate internal menus related to the menus, input windows and graphics windows. They are as follows:

DIR

This has been discussed earlier under the FILES menu.

EXPORT

This function too has already been discussed under the FILES menu. Its use is slightly different in this context. For example, if bivariate data are exported, the only detail to be supplied is the name of the output file, plus possibly some comments.

PRINT

This prints the contents of the current window. If the window does not contain graphics (for example, it is an input or help window) printing is only possible for users who have an Epson-compatible dot matrix printer. It is also important to note that DMC asks the user for the name of a file as the output destination. Actually, printing may be directed to a printer (using a name reserved in DOS (PRN, LPT1, LPT2, ...)) as well as to a file which can be sent to the printer later on. It should be remembered that DMC can use only parallel ports as physical output channels, and it is therefore necessary to redirect the serial printers using DOS commands. For further information see the PRINTER command in the OPTS menu.

Macintosh. On Macintosh machines this function is available for printing plots only.

SAVE

This function allows the user to direct the graphics produced to files (whose names the user must specify). DMC automatically places these files in the appropriate work directory (PLOT). They can be retrieved and eventually printed by means of the LOAD PLOT command in the FILES menu. Note that DMC automatically adds the 'DMP' extension to these files and it is therefore unnecessary to specify an extension.

VIRTUAL

This function is analogous to the SAVE function. It displays the current graphics on a 'virtual' screen (having a resolution greater than the standard one and equal to 712×712 pixels) and saves it on the file specified by the user.

Macintosh. This function is not supported.

COORDS

This function allows the user to obtain the co-ordinates of any point of the screen by simply indicating the point with the cursor and pressing the left mouse key.

Keyboard users may use the following commands:

- <PGUP>: increases the speed of cursor movement;
- <PGDN>: decreases the speed of cursor movement;
- cursor keys: move the cursor;
- <RETURN>: requests the values of the co-ordinates of the point indicated by the cursor;

In each case the user returns to the internal menu by pressing the

<ESC> key. Note that only an approximate evaluation of co-ordinates is performed, due to the scale reduction necessary to portray the graphics on the screen.

ZOOM

This function can be used to enlarge previously selected zones of the current graphics. The selection may be done in two different ways:

- With the mouse: DMC displays a rectangle which may be shifted by moving the mouse. The dimensions of the rectangle can be altered by using the <+> and <–> keys, thus obtaining, respectively, an enlargement or reduction of its dimensions. Once the positioning operation is complete, the left key of the mouse is pressed.
- Using the keyboard: in this case the command keys are as follows:
- cursor keys: move the rectangle into the desired position;
- <+>: increases the rectangle's dimensions;
- <–>: decreases the rectangle's dimensions;
- <PGUP>: increases the speed of the rectangle's movement;
- <PGDN>: decreases the speed of the rectangle's movement;
- <RETURN>: establishes the final position of the rectangle.

In all cases, the user may return to the internal menu by pressing the <ESC> key. This function cannot be applied to three-dimensional plots and, in general, to those for which the error message 'Function not available' is displayed.

←

This function lets the user exit from the internal menus. Selecting this is the same as using the <ESC> key. It is convenient to use this function if a mouse is used (the internal menus do not have an exit zone).

SELECT

The file whose name is displayed at the top of the current window can be selected by using this function in order to perform the following:

- examine its type with the FILE INFO command;
- display the contents with the external command activated by the LIST function;
- delete it by using the ERASE function.

It can also be selected in a more immediate way by using the mouse or by pressing the <SHIFT> key.

Macintosh. This function is superfluous.

FILE INFO

If the file is of a binary type and is DMC-produced, this function allows its nature to be determined. For example, it tells the user which model has generated the data, how many iterations have been computed, etc.

LIST

This function has been created to recall an external program, which may be an editor or any other type of program which allows the user to examine the contents of a selected file. The name of the external program is contained in the CONFIG.DMC file. By default this recalls the FLIST.BAT batch program which uses the DOS 'MORE' command.

Macintosh. This function is not available because the Macintosh operating system does not support the 'command line'.

ERASE

The file selected with the SELECT function is deleted if confirmed by the user.

Finally, there is a 'floating' menu with which DMC assures itself of the user's intention to delete a pre-existing file by copying new data over it. It makes use of the following two commands:

REPLACE

DMC has revealed that the file indicated as output file for the current function already exists. This function allows it to be used again.

CANCEL

This option avoids destroying the pre-existing file.

15.5 DMC INTERNAL COMPILER

DMC is equipped with an internal compiler which translates dynamic models into machine language. The correct syntax to follow is intuitive and does not require any additional effort for those who know such programming languages as BASIC, Pascal or C. Of course, the machine's immediate compilation of the equations of dynamic models into a directly executable code allows the maximum possible computational efficiency. The functions available in DMC follow, along with their evaluation priority. Remember that low values indicate high priorities.

=   Priority 8

Assignment operator. Transfers the variable or the result of the expression on the right of the symbol to the variable on the left.
Example:

eq1 = x, x=10 → eq1 = 10

+   Priority 3

Summation operator. Returns the sum of two operands.
Example:

x + y,
x=10, y=2 → 10+2 → 12

−   Priority 3

Subtraction operator. Returns the difference between the first and second operands.
Example:

x−y,
x=10, y=2 → 10−2 → 8

−   Priority 1

Sign change operator. Changes the sign of the expression which follows it.
Example:

−x, x=10 → −10

*   Priority 2

Multiplication operator. Returns the product of two operands.
Example:

x * y,
x=4, y=6 → 4 * 6 → 24

/   Priority 2

Division operator. Returns the quotient of the first operand divided by the second.
Example:

x/y,
x=10, y=2 → 10/2 → 5

^   Priority 1

Power operator. Returns the value of the first operand raised to the power indicated by the second.
Example:

x^y,
x=2, y=4 → 2^4 = 16

%   **Priority 2**

Modulus operator. Returns the remainder of the division of the first operand by the second.
Example:

x%y,
x=1, y=0.4 → 1 % 0.4 → 0.2

>   **Priority 4**

Comparison operator. If the first operand is greater than the second, it produces 1.0 as a result, otherwise 0.0.
Example:

x > y,
x=10, y=2 → 10 > 2 → 1.0
x=2, y=10 → 2 > 10 → 0.0

<   **Priority 4**

Comparison operator. If the first operand is smaller than the second, it produces 1.0 as a result, otherwise 0.0.
Example:

x < y,
x=10, y=2 → 10 < 2 → 0.0
x=2, y=10 → 2 < 10 → 1.0

>=   **Priority 4**

Comparison operator. If the first operand is greater or equal to the second, it produces 1.0 as a result, otherwise 0.0.
Example:

x >= y,
x=10, y=2 → 10 >= 2 → 1.0
x=10, y=10 → 10 >= 10 → 1.0

<=   **Priority 4**

Comparison operator. If the first operand is smaller than or equal to the second, it produces 1.0 as a result, otherwise 0.0.
Example:

```
x <= y,
x=10, y=2 → 10 <= 2 → 0.0
x=10, y=10 → 10 <= 10 → 1.0
```

==   Priority 5

Comparison operator. If the first operand is equal to the second, it produces 1.0 as a result, otherwise 0.0.
Example:

```
x == y,
x=10, y=2 → 10 == 2 → 0.0
x=10, y=10 → 10 == 10 → 1.0
```

!=   Priority 5

Comparison operator. If the first operand is different from the second, it produces 1.0 as a result, otherwise 0.0.
Example:

```
x != y,
x=10, y=2 → 10 != 2 → 1.0
x=10, y=10 → 10 != 10 → 0.0
```

&&   Priority 6

Boolean operator. If the operands are different from zero, it produces 1.0, otherwise 0.0.
Example:

```
x && y,
x=1, y=8 → 1 && 8 → 1.0
x=1, y=0 → 1 && 0 → 0.0
```

|| Priority 7

Boolean operator. If at least one operand is different from zero it produces 1.0, otherwise 0.0.
Example:

```
x || y,
x=1, y=8 → 1 || 8 → 1.0
x=1, y=0 → 1 || 0 → 1.0
x=0, y=0 → 0 || 0 → 0.0
```

fact(*n*) Priority 1

Returns the factorial of the number *n*. Note that this function accepts integer arguments only and, therefore, ignores the decimal part of *n* which may be present.
Example:

fact(7) = $7 \cdot 6 \cdot 5 \cdot 4 \cdot 3 \cdot 2 \cdot 1 = 5040$

pi(*n*) Priority 1

Pi function. Returns the value of π divided by n.
Example:

pi(4) = 0.7853...
pi(2) = 1.5707...
pi(1) = 3.1415...

sin(α) Priority 1

Sine function. Returns the sine of the angle (in radians) in parentheses.
Example:

sin(x), x=0 $\rightarrow$ sin(0) $\rightarrow$ 0.0

cos(α) Priority 1

Cosine function. Returns the cosine of the angle (in radians) in parentheses.
Example:

cos(x), x=0 $\rightarrow$ cos(0) $\rightarrow$ 1.0

tan(α) Priority 1

Tangent function. Returns the tangent of the angle (in radians) in parentheses.
Example:

tan(pi/4), pi=3.1415... $\rightarrow$ 1.0

asin($\cdot$) Priority 1

Arc-sine function. Returns the angle whose sine is indicated in parentheses.
Example:

asin(1) $\rightarrow$ pi/2, pi=3.1415...

acos(·)   Priority 1

Arc-cosine function. Returns the angle whose cosine is indicated in parentheses.
Example:

  acos(1) → 0

atan(·)   Priority 1

Arc-tangent function. Computes the angle whose tangent is indicated in parentheses.
Example:

  atan(1) → pi/4, pi=3.1415...

exp(·)   Priority 1

Returns the value of e (=2.7182818...) raised to the power indicated in parentheses.
Example:

  exp(1) → 2.7182818...

log(·)   Priority 1

Returns the logarithm to base e (=2.7182818...) of the number indicated in parentheses.
Example:

  log(e) → 1

lg10(·)   Priority 1

Returns the logarithm, to base 10 of the number indicated in parentheses.
Example:

  lg10(100) → 2

abs(·)   Priority 1

Returns the absolute value of the number indicated in parentheses.
Example:

  abs(−100) → 100
  abs(100) → 100

rndr(β)   Priority 1

Generates a random number from a uniform distribution on [0,β].

rndn(μ) Priority 1

Generates a random number from a normal distribution $\mathcal{N}(\mu, 1)$.

rndg(α) Priority 1

Generates a random number from a gamma distribution with parameter α.

test($\cdot$)
iftrue($\cdot$)
iffalse($\cdot$)

These functions allow the user to assign a value to a variable if a condition is met.
Example:

    a = 1
    x = 10
    y = 20
    q = test(a>10) iftrue(x) iffalse(x+y) → q = 30
    p = test(a<10) iftrue(x^2) iffalse(y^2) → p = 100

The priority of operators and functions may be varied by the use of parentheses. For example, in the expression:

    3 * (x + y)

the operation of the sum, between parentheses, is evaluated before the multiplication operation even though it has a lower priority.

References

Abraham, R.H. and Marsden, J.E. (1978). *Foundations of Mechanics*. Reading, Mass.: Benjamin and Cummings.

Abraham, R.H. and Shaw, C.D. (1988). *Dynamics, the Geometry of Behaviour*, Parts I and II. Santa Cruz: Aerial.

Anderson, P.W., Arrow, K. and Pines, D. (eds.) (1988). *The Economy as an Evolving Complex System*. Santa Fe Institute Studies in the Sciences of Complexity: Addison–Wesley.

Anosov, D.V. (1967). Geodesic flows and closed Riemannian manifolds with negative curvature. *Proc. Steklov Inst. Mat.* **90**.

Arneodo, A., Coullet, P. and Tresser, C. (1981). Possible new strange attractors with spiral structure. *Communications in Mathematical Physics*, **79**, 573–579.

Arneodo, A., Coullet, P. and Tresser, C. (1982). Oscillators with chaotic behaviour: an illustration of a theorem by Šilnikov. *Journal of Statistical Physics*, **27**, 171–182.

Arrow, K.J. (1988). Workshop on the economy as an evolving complex system: Summary. In Anderson, Arrow and Pines (eds.) 275–282.

Bak, P., Bohr, T. and Jensen, M.H. (1984). Mode-locking and the transition to chaos in dissipative systems. *Phys. Scr.*, **9T**, 50–58.

Barnett, W. (1980). Economic monetary aggregates: an application of index number and aggregation theory. *Journal of Econometrics*, **14**, 11–48.

Barnett, W. and Chen, P. (1986). The aggregation theoretic monetary aggregates are chaotic and have strange attractors. In Barnett, W., Berndt, E. and White, H. (eds.), *Dynamic Econometric Modelling*, 199–246. Cambridge: Cambridge University Press.

Barnett, W. and Choi, S.S (1989). A Monte Carlo study of tests of blockwise weak separability. In Barnett, W., Gewerke, J. and Shell K. (eds.), *Economic Complexity: Chaos, Sunspots, Bubbles and Nonlinearity*, 141–213. Cambridge: Cambridge University Press.

Baumol, W.J. and Benhabib, J. (1989). Chaos: significance, mechanism, and economic applications. *Journal of Economic Perspectives*, **3**, 77–107.

Benedicks, M. and Carleson, L. (1989). The dynamics of the Hénon map. Preprint.

Benettin, G. and Galgani, L. (1979). Lyapunov characteristic exponents and stochasticity. In Laval G. and Grésillon, D. (eds.), *Intrinsic Stochasticity in Plasmas*, 93–114. Orsay: Editions de Physique.

Benettin, G., Galgani, L., Giorgilli, A. and Strelcyn, J.M. (1980). Lyapunov characteristic exponents for smooth dynamical systems; a method for computing them all. *Meccanica*, **15**, 21–30.

Benhabib, J. and Day, R.H. (1981). Rational choice and erratic behaviour. *Review of Economic Studies*, **48**, 459–471.

Benhabib, J. and Day, R.H. (1982). A characterization of erratic dynamics in the overlapping generations models. *Journal of Economic Dynamics and Control*, **4**, 37–55.

Benhabib, J. and Nishimura, K. (1979). The Hopf bifurcation and the existence and stability of closed orbits in multisector models of optimal economic growth. *Journal of Economic Theory*, **21**, 417–444.

Bergé, P., Pomeau, Y. and Vidal, C. (1984). *Order within Chaos*. New York: Wiley.

Billingsley, P. (1965). *Ergodic Theory and Information*. New York: Wiley.

Billingsley, P. (1979). *Probability and Measure*. New York: Wiley.

Blanchard, O.J. and Fischer, S. (1987). *Lectures on Mathematics*. Cambridge, Mass.: MIT Press.

Blinder, A.S. (1989). *Macroeconomics under Debate*. New York: Harvester Wheatsheaf.

Boldrin, M. and Montrucchio, L. (1986). On the indeterminacy of capital accumulation paths. *Journal of Economic Theory*, **40**, 26–39.

Boldrin, M. and Woodford, M. (1990). Equilibrium models displaying endogenous fluctuations and chaos: a survey. *Journal of Monetary Economics*, **25**, 189–223.

Bowen, R. (1978). On axiom A diffeomorphisms. *CBNS Regional Conference Series in Mathematics*, **35**. Providence: ANS Publications.

Brock, W.A. (1986). Distinguishing random and deterministic systems: abridged version. *Journal of Economic Theory*, **40**, 168–195.

Brock, W.A. (1988). Nonlinearity and complex dynamics in economics and finance. In Anderson, Arrow and Pines (eds.), 77–97.

Brock, W.A., Dechert, W.D. and Scheinkmann, J. (1987). A test for independence based on the correlation dimension. SSRI Working Paper No. 8702, University of Wisconsin.

Brock, W.A. and Sayers, C.L. (1988). Is the business cycle characterized by deterministic chaos? *Journal of Monetary Economics*, **22**, 71–90.

Broomhead, D.S., Jones, R. and King, G.P. (1987). Topological dimension and local co-ordinates from time series data. *Journal of Physics*, **A 20**, L563.

Broomhead, D.S. and King, G.P. (1986a). Extracting qualitative dynamics from experimental data. *Physica*, **20D**, 217–236.

Broomhead, D.S. and King, G.P. (1986b). On the qualitative analysis of experimental dynamical systems. In Sarkar, S. (ed.), *Nonlinear Phenomena and Chaos*, 113–145. Bristol: Adam Hilger.

Burns, A.F. and Mitchell, W.C. (1946). *Measuring Business Cycles*. New York: National Bureau of Economic Research.

Carleson, L. (1989). *Stochastic Behaviour of Deterministic Systems*. Working Paper No. 233, The Industrial Institute for Economic and Social Research, Stockholm.

Chen, P. (1988a). Empirical and theoretical evidence of economic chaos. *System Dynamics Review*, **4**, 81–108.

Chen, P. (1988b). Multiperiodicity and irregularity in growth cycles: a continuous model of monetary attractors. *Mathematical Computing and Modelling*, **10**, 647–660.

Cohen, D.L. (1980). *Measure Theory*. Boston: Birkhäuser.

Collet, P. and Eckmann, J.P. (1980). *Iterated Maps on the Interval as Dynamical Systems*. Basle: Birkhäuser.

Collet, P., Eckmann, J.P. and Lanford, O.E. (1980). Universal properties of maps on an interval. *Communications of Mathematical Physics*, **76**, 211–254.

Crutchfield, J.P., Farmer, J.D. and Huberman, B.A. (1982). Fluctuations and simple chaotic dynamics. *Physics Reports*, **92**, 45–82.

Crutchfield, J.P. and McNamara, B.S. (1987). Equations of motions from a data series. *Complex Systems*, **1**, 417–452.

Cvitanovic, P. (ed.) (1984). *Universality in Chaos*. Bristol: Adam Hilger.

Day, R.H. (1982). Irregular growth cycles. *American Economic Review*, **72**, 406–414.

Deneckere, R. and Pelikan, S. (1986). Competitive chaos. *Journal of Economic Theory*, **40**, 13–25.

Devaney, R.L. (1986). *An Introduction to Chaotic Dynamical Systems*. Menlo Park, Calif.: Benjamin and Cummings.

Diamond, P. (1976). Chaotic behaviour of systems of difference equations. *International Journal of Systems Science*, **6**, 953–956.

Doetsch, G. (1971). *Guide to the Applications of the Laplace and Z–Transform*. London: Van Nostrand Reinhold.

Eckmann, J.P. (1981). Roads to turbulence in dissipative dynamical systems. *Review of Modern Physics*, **53**, 643–654.

Eckmann, J.P. and Ruelle, D. (1985). Ergodic theory of chaos and strange attractors. *Review of Modern Physics*, **57**, 617–656.

Farmer, J.D. and Sidorowich, J.J. (1988). Exploiting chaos to predict the future and reduce noise. In Lee Y.C. (ed.), *Evolution, Learning and Cognition*. World Scientific.

Feigenbaum, M.J. (1978). Quantitative universality for a class of nonlinear transformations. *Journal of Statistical Physics*, **19**, 25–52.

Foley, D. (1975). On two specifications of asset equilibrium in macroeconomic models. *Journal of Political Economy*, **83**, 303–324.

Ford, J. (1983). How random is a coin toss? *Physics Today*, **36**, 40–48.

Franceschini, V. and Tebaldi, C. (1979). Sequences of infinite bifurcations and turbulence in a five–mode truncation of the Navier–Stokes equations. *Journal of Statistical Physics*, **21**, 707.

Frank, M.Z. and Stengos, T. (1988). Some evidence concerning macroeconomic chaos. *Journal of Monetary Economics*, **22**, 423–438.

Frank, M.Z. and Stengos, T. (1989). Measuring the strangeness of gold and silver rates of return. *Review of Economic Studies*, **56**, 553–567.

Frisch, R. (1933). Propagation problems and impulse problems in dynamic economics. In *Economic Essays in Honour of Gustav Cassel*, 171–205. London: Allen and Unwin.

Gandolfo, G. (1983). *Economic Dynamics: Methods and Models*, 2nd edition. Amsterdam: North Holland.

Gleick, J. (1987). *Chaos – Making a New Science*. New York: Viking.

Goodwin, R.M. (1951). The nonlinear accelerator and the persistence of business cycles. *Econometrica*, **19**, 1–17.

Goodwin, R.M. (1967). A growth cycle. In Feinstein, C.H. (ed.), *Socialism, Capitalism and Economic Growth*, 54–58. Cambridge: Cambridge University Press. Revised version in Hunt, E.K. and Schwarz, J.G. (eds.) (1969). *A Critique of Economic Theory*, 442–449. Harmondsworth: Penguin.

Grandmont, J.M. (1985). On endogenous competitive business cycles. *Econometrica*, **53**, 995–1045.

Granger, C.W.J. (ed.) (1990). *Modelling Economic Series*. Oxford: Oxford University Press.

Granger, C.W.J. and Newbold, P. (1977). *Forecasting Economic Time Series*. New York: Academic Press.

Grassberger, P. and Procaccia, I. (1983). Measuring the strangeness of strange attractors. *Physica*, **9D**, 189–208.

Grebogi, C., Ott, E., Pelikan, S. and Yorke, J.A. (1984). Strange attractors that are not chaotic, *Physica*, **13D**, 261–268.

Grebogi, C., Ott, E. and Yorke, J.A. (1983). Chaotic attractors in crisis. *Physics Review Letters*, **48**, 1510–1597.

Guckenheimer, J. (1977). On the bifurcation of maps of the interval. *Inventiones Mathematicae*, **39**, 165–178.

Guckenheimer, J. and Holmes, P. (1983). *Nonlinear Oscillations, Dynamical Systems and Bifurcations of Vector Fields*. New York: Springer Verlag.

Hadamard, J. (1898). Sur l'itération et les solutions asymptotiques des équations différentials. *Bull. Soc. Mat. France*, **29**, 224–228.

Hammel, S.M., Yorke, J.A. and Grebogi, C. (1987). Do numerical orbits of chaotic dynamical processes represent true orbits? *Journal of Complexity*, 3, 136–145.

Hao, B.L. (1984). *Chaos*. Singapore: World Scientific.

Hao, B.L. (1989). *Elementary Symbolic Dynamics and Chaos in Dissipative Systems*. Singapore: World Scientific.

Hartman, P. (1964). *Ordinary Differential Equations*. New York: Wiley.

Hassard, B.D., Kazarinoff, N.D., Wan, Y.-H. (1980). *Theory and Applications of the Hopf Bifurcation*. Cambridge: Cambridge University Press.

Helleman, R.H.G. (1980). Self-generated chaotic behaviour in nonlinear mechanics. In Cohen, E.G.D. (ed.), *Fundamental Problems in Statistical Mechanics*, vol. V, Amsterdam: North Holland. Reprinted in Cvitanovic (ed.) (1984), 420–488.

Hicks, J.R. (1950). *A Contribution to the Theory of the Trade Cycle*. Oxford: Oxford University Press.

Holden, A.V. (ed.) (1986). *Chaos*. Manchester: Manchester University Press.

Holmes, P.J. and Whiteley, D.C. (1984). Bifurcations of one– and two–dimensional maps. *Philosophical Transactions of the Royal Society, London*, **A311**, 43–102; erratum, *ibid.*, 601.

Hopf, E. (1942). Abzweigungen einer periodischen Lösung von einer stationären Lösung eines Differentialgleichungssystems. *Math. Naturwiss. Klasse*. Leipzig: Söchs. Akademie der Wissenschaften, **94**, 1.

Invernizzi, S. and Medio, A. (1991). On lags and chaos in economic dynamic models. *Journal of Mathematical Economics*, **20**, 521–550.

Jackson, E.A. (1989). *Perspectives on nonlinear dynamics*, 2 vols. Cambridge: Cambridge University Press.

Kaldor, N. (1940). A model of the trade cycle. *Economic Journal*, **50**, 78–92.

Kaplan, W. (1962). *Operational Methods for Linear Systems*. Reading, Mass.: Addison–Wesley.

Kaplan, J.L. and Yorke, J.A. (1979). Chaotic behavior of multidimensional difference equations. In Peitgen, H.O. and Walther, H.O. (eds.), *Functional Differential Equations and Approximation of Fixed Points* (Lecture Notes in Mathematics, **730**). New York: Springer Verlag.

Katok, A.B. (1980). Lyapunov exponents, entropy and periodic points for diffeomorphisms. *Publ. Math. IHES*, **51**, 137–174.

Kelsey, D. (1988). The economics of chaos or the chaos of economics. *Oxford Economic Papers*, **40**, 1–31.

Lanford, O.E. (1982). A computer assisted proof of the Feigenbaum conjectures. *Bulletin of the American Mathematical Society*, **6**, 427–434.

Lauwerier, H.A. (1986). One-dimensional iterative maps. In Holden (ed.), 39–57.

Lax, P.D. (1986). Mathematics and computing. *Journal of Statistical Physics*, **43**, 749–756.

Li, T.Y. and Yorke, J.A. (1975). Period three implies chaos. *American Mathematical Monthly*, **82**, 985–992.

Lichtenberg, A.J. and Lieberman, M.A. (1983). *Regular and Stochastic Motion*. New York: Springer Verlag.

Lines, M. (1991). Slutzky and Lucas: random causes of the business cycle. *Structural Change and Economic Dynamics* (forthcoming).

Lorenz, E.N. (1963). Deterministic non-periodic flow. *Journal of Atmospheric Science*, **20**, 130–141.

Lorenz, E.N. (1980). Noisy periodicity and reverse bifurcation. *Annals of the New York Academy of Sciences*, **357**, 282–292.

Lorenz, E.N. (1984). The local structure of a chaotic attractor in four dimensions. *Physica*, **13D**, 90–104.

Lorenz, H.W. (1988). Spiral-type attractors in low-dimensional continuous-time dynamical systems. University of Göttingen. Mimeo.

Lorenz, H.W. (1989a). Strange attractors in dynamical economics. In Ames, W.F. and Brezinski, C. (eds.), *IMACS Transactions on Scientific Computing*, vol. I. Basle: J.C. Balzer.

Lorenz, H.W. (1989b). *Nonlinear Dynamical Economics and Chaotic Motion*. New York: Springer Verlag.

Lyapunov, A.M. (1907). Problème général de stabilité de mouvement (translation from Russian). *Annuaire de la Faculté des Sciences, Université de Toulouse*, **9**, 203–475. Reproduced in *Annals of Mathematical Study*, **17**. Princeton.

McCawley, J.L., Jr and Palmare, J.I. (1986). Computable chaotic orbits. *Physics Letters* A.

Mañé, R. (1981). On the dimension of the compact invariant sets of certain nonlinear maps. In Rand, D.A. and Young, L.S. (eds.), *Dynamical Systems and*

Turbulence. (Lecture Notes in Mathematics, **898**), 230–242. New York: Springer Verlag.

Marotto, F.R. (1978). Snap-back repellers imply chaos in R^n. *Journal of Mathematical Analysis and Application*, **63**, 199–223.

May, J. (1970). Period analysis and continuous analysis in Patinkin's macroeconomic model. *Journal of Economic Theory*, **2**, 1–9.

May, R. M. (1976). Simple mathematical models with very complicated dynamics. *Nature*, **261**, 459–467.

May, R.M. (1983). Nonlinear problems in ecology and resource management. In Ioos, G., Helleman, R.H.G. and Stora, R. (eds.), *Chaotic Behaviour of Deterministic Systems*, 514–563. Amsterdam: North Holland.

Medio, A. (1991a). Discrete and continuous-time models of chaotic dynamics in economics, *Structural Change and Economic Dynamics*, **2**, 99–118.

Medio, A. (1991b). Continuous-time models of chaotic dynamics in economics. *Journal of Economic Behavior and Organization*, **16**, 115–151.

Melnikov, V.K. (1963). On the stability of the center for time periodic perturbations. *Transactions of the Moscow Mathematical Society*, **12** (1), 1–57.

Metzler, L.A. (1941). The nature and stability of inventory cycles. *Review of Economic Studies*, **23**, 113–129.

Moon, F.C. (1987). *Chaotic Vibrations.* New York: Wiley.

Moon, F.C. and Holmes, W.T. (1985). Double Poincaré sections of a quasi periodically forced, chaotic attractor. *Physics Letters*, **111A** (4), 157–160.

Moser, J. (1973). *Stable and Random Motions in Dynamical Systems.* Princeton: Princeton University Press.

Murata Y. (1977). *Mathematics for Stability and Optimization of Economic Systems* New York: Academic Press.

Muth, J. (1961). Rational expectations and the theory of price movements. *Econometrica*, **29**, 315–335.

Newhouse, S., Ruelle, D. and Takens, F. (1978). Occurrence of strange axiom A attractors near quasi periodic flows on T^3, $m > 3$. *Communications in Mathematical Physics*, **64**, 35–40.

Oono, Y. and Osikawa, M. (1980). Chaos in nonlinear difference equations. *Progress in Theoretical Physics*, **64**, 54–67.

Oseledec, V.I. (1968). A multiplicative ergodic theorem. Lyapunov characteristic numbers for dynamical systems. *Transactions of the Moscow Mathematical Society*, **19**, 197–231.

Parker, T.S. and Chua, L.O. (1987). Chaos: a tutorial for engineers. *Proc. IEEE*, **75**, 982–1007.

Parker, T.S. and Chua, L.O. (1989). *Practical Numerical Algorithms for Chaotic Systems.* New York: Springer Verlag.

Peixoto, M.M. (1962). Structural stability on two-dimensional manifolds. *Topology*, **1**, 101–120 .

Press, W.H., Flannery, B.P., Tenkolsky, S.A. and Vetterling, W.T. (1986). *Numerical Recipes. The Art of Scientific Computing.* Cambridge: Cambridge University Press.

Priestley, M.B. (1981). *Spectral Analysis and Time Series*, 2 vols. London: Academic Press.

Puu, T. (1987). Complex dynamics in continuous models of the business cycle. *Lecture Notes in Economic and Mathematical Systems*, **293**, 227. Berlin: Springer Verlag.

Ramsey, J.B. and Yuan, H. (1989). Bias and error bars in dimension calculation and their evaluation in some simple models. *Physics Letters*, **134A**, 287–297.

Ramsey, J.B., Sayers, C.L. and Rothman, P. (1990). The statistical properties of dimension calculations using small data sets: some economic applications. *International Economic Review*, **31**, 991–1020.

Reichlin, P. (1986). Equilibrium cycles in an overlapping generations economy with production. *Journal of Economic Theory*, **40**, 89–102.

Robinson, C. (1988). Bifurcation to a transitive attractor of Lorenz type. Preprint.

Rössler, O.E. (1976). Chemical turbulence: chaos in a small reaction-diffusion system. *Z. Naturforsch.*, **31A**, 1168–1172.

Ruelle, D. (1989). *Chaotic Evolution and Strange Attractors*. Cambridge: Cambridge University Press.

Ruelle, D. and Takens, F. (1971). On the nature of turbulence. *Communications of Mathematical Physics*, **20**, 167–192.

Ryder, H.E., Jr. and Heal, G.M. (1973). Optimal growth with intertemporary dependent preferences. *Review of Economic Studies*, **1**, 1–31.

Sarkowskii, A.N. (1964). Coexistence of the cycles of a continuous mapping of the line into itself. *Ukr. Mat. Zh.*, **16 (1)**, 61–71.

Sauer, T., Yorke, J.A., Casdagli, M. and Kostelich, E. (1990). Embedology. University of Maryland, Mimeo.

Scheinkman, J.A. (1990). Nonlinearities in economic dynamics. *Economic Journal*, **100**, 33–49.

Scheinkman, J.A. and LeBaron, B. (1989a). Nonlinear dynamics and stock returns. *Journal of Business*, **62**, 311–337.

Scheinkman, J.A. and LeBaron, B. (1989b). Nonlinear dynamics and GNP data. In Burnett, W., Gewerke, J. and Shell, K. (eds.), *Economic Complexity: Chaos, Sunspots, Bubbles and Nonlinearity*, 213–231. Cambridge: Cambridge University Press.

Schuster, H.G. (1989). *Deterministic Chaos: An Introduction*, 2nd edition. VCH: Weinheim.

Shannon, C.E. (1948). A mathematical theory of communication. *Bell System Tech. J.*, **27**, 379–423; 623–656.

Shaw, R. (1980) Strange attractors, chaotic behaviour, and information flow. *Z. Naturforsch.*, **36A**, 80–112.

Šilnikov, L.P. (1965). A case of the existence of a denumerable set of periodic motions. *Sov. Math. Dokl.*, **6**, 163. Russian original, *Doklady*, **189**, 588.

Šilnikov, L.P. (1970). On a new type of bifurcation of multidimensional dynamical systems. *Sov. Math. Dokl.*, **10**, 1368. Russian original, *Doklady*, **189**, 59.

Simò, C. (1979). On the Hénon–Pomeau attractor. *Journal of Statistical Physics*, **21**, 465.

Sirovich, L. (1988). *Introduction to Applied Mathematics*. New York: Springer Verlag.

Slutsky, E. (1927). The summation of random causes as the source of cyclical processes. *Econometrica*, **5**, 105–146.

Smale, S. (1963). Diffeomorphisms with many periodic points. In Cairns, S.S. (ed.), *Differential and Combinatorial Topology*, 63–80. Princeton: Princeton University Press.

Smale, S. (1967). Differential dynamical systems. *Bulletin of the American Mathematical Society*, **73**, 747–817.

Sparrow, C. (1980). Bifurcation and chaotic behaviour in simple feedback systems. *Journal of Theoretical Biology*, **83**, 93–105.

Sparrow, C. (1982). *The Lorenz Equations*. New York: Springer Verlag.

Sparrow, C. (1986). The Lorenz equations. In Holden (ed.), 111–135.

Sraffa, P. (1960). *Production of Commodities by Means of Commodities*. Cambridge: Cambridge University Press.

Stutzer, M.J. (1980). Chaotic dynamics and bifurcation in a macro-model. *Journal of Economic Dynamics and Control*, **2**, 353–376.

Takens, F. (1981). Detecting strange attractors in turbulence. In Rand, D.A. and Young, L.S. (eds.), *Dynamical Systems and Turbulence*, 366–381. New York: Springer Verlag.

Temam, R. (1988). *Infinite-Dimensional Dynamical Systems in Mechanics and Physics*. New York: Springer Verlag.

Theil, H. and Fiebig, D. (1981). A maximum entropy approach to the specification of distributed lags. *Economic Letters*, **7**, 339–342.

Theil, H. and Fiebig, D. (1984). *Exploring Continuity: Maximum Entropy Estimation of Continuous Distribution*. Cambridge, Mass.: Ballinger.

Thom, R. (1975). *Structural Stability and Morphogenesis*. Reading, Mass.: W.A. Benjamin.

Thompson, J.M.T. and Stewart, H.B. (1986). *Nonlinear Dynamics and Chaos*. New York: Wiley.

Tomita, K. (1986). Periodically forced nonlinear oscillators. In Holden (ed.), 211–236.

Turnovsky, S.J. (1977). On the formulation of continuous time macroeconomic models with asset accumulation. *International Economic Review*, **18**, 1–28.

Ulam, S.M. and von Neumann, J. (1947). On combination of stochastic and deterministic processes. *Bulletin of the American Mathematical Society*, **53**, 1120.

Vautard, R. and Ghil, M. (1989). Singular spectrum analysis in nonlinear dynamics with applications to paleoclimatic time series. *Physica*, **35B**, 395–424.

Whitley, D. (1983). Discrete dynamical systems in dimensions one and two. *Bulletin of the London Mathematical Society*, **15**, 177–217.

Wiggins, S. (1988). *Global Bifurcations and Chaos: Analytical Methods*. New York: Springer Verlag.

Wiggins, S. (1990). *Introduction to Applied Nonlinear Dynamical Systems and Chaos*. New York: Springer Verlag.

Wolf, A. and Swift, J.B. (1984). Progress in computing Lyapunov exponents from experimental data. In Horton, C.W., Jr. and Reichlin, L.E. (eds.), *Statistical Physics and Chaos in Fusion Plasma*. New York: Wiley.

Wolf, A., Swift, J.B., Swinney, H.L. and Vastano, J.A. (1985). Determining Lyapunov exponents from a time series. *Physica*, **16D**, 285–317.

Woodford, M. (1987). *Equilibrium Models of Endogenous Fluctuations*. University of Chicago. Mimeo.

Yoshida, K. (1984). *Operational Calculus: A Theory of Hyperfunctions*. New York: Springer Verlag.

Yule, G.U. (1927). On a method of investigating periodicities in disturbed series. *Philosophical Transactions*, **226A**, 267–298.

Index